INTRODUCTION TO BRYOPHYTES

Bryophytes mark a pivotal step in land plant evolution, and their significance in the regulation of ecosystems and the conservation of biodiversity is becoming increasingly acknowledged. This introductory textbook assumes no prior knowledge of bryophyte biology, making it ideal for advanced undergraduate and graduate students, as well as for amateur botanists. The authors expertly summarize the diversity of bryophytes and outline recent advances in our understanding of their evolutionary history, their ecological roles and preferences, their distribution patterns and conservation needs. The text is highly illustrated throughout, with boxed summaries of topics of current relevance in bryophyte biology, and a glossary of technical terms.

ALAIN VANDERPOORTEN is a Research Associate of the Belgian Funds for Scientific Research at the University of Liège, Belgium, where he teaches molecular systematics, biogeography and landscape ecology. His research presently focuses on the evolution of endemism.

BERNARD GOFFINET is an Associate Professor in Ecology and Evolutionary Biology at the University of Connecticut. His current research spans chloroplast genome evolution in liverworts, the phylogenetic history of mosses and the systematics of lichen-forming fungi. He teaches bryophyte and lichen biology and green plant evolution.

INTRODUCTION TO BRYOPHYTES

ALAIN VANDERPOORTEN
University of Liège, Belgium

and

BERNARD GOFFINET
University of Connecticut, USA

CAMBRIDGE
UNIVERSITY PRESS

CAMBRIDGE UNIVERSITY PRESS

Cambridge, New York, Melbourne, Madrid, Cape Town, Singapore, São Paulo,
Delhi, Dubai, Tokyo, Mexico City

Cambridge University Press
The Edinburgh Building, Cambridge CB2 8RU, UK

Published in the United States of America by Cambridge University Press, New York

www.cambridge.org
Information on this title: www.cambridge.org/9780521877121

© A. Vanderpoorten and B. Goffinet 2009

First published 2009
Reprinted 2010

Printed in the United Kingdom at the University Press, Cambridge

A catalogue record for this publication is available from the British Library

Library of Congress Cataloging-in-Publication Data

Introduction to bryophytes / edited by Alain Vanderpoorten and Bernard Goffinet.
p. cm.
Includes bibliographical references and index.
ISBN 978-0-521-87712-1 (hardback) – ISBN 978-0-521-70073-3 (pbk.)   1. Bryophytes.
I. Vanderpoorten, Alain.   II. Goffinet, Bernard.   III. Title.

QK533.I57 2009
588–dc22

2008054301

ISBN 978-0-521-87712-1 Hardback
ISBN 978-0-521-70073-3 Paperback

Contents

Preface

The concept for a book providing an introduction to the biology of bryophytes emanates from our passion for these land plants and the lack of recent manuals offering general insights into their fascinating diversity and evolutionary history. Throughout its history, bryological research has contributed significantly to the field of plant sciences, starting with the discovery of sex chromosomes in plants and culminating most recently with the assembly of the *Physcomitrella* genome, fundamental for the study of the evolution of genes and their function during the diversification of land plants. Bryophytes are, in fact, pivotal in land plant evolution, and the recent advances in molecular phylogenetics and genomics have allowed for a clearer picture of land colonization and subsequent evolution to emerge. Ecologically, the significance of bryophytes for the regulation of global biogeochemical cycles, especially carbon, has long been acknowledged and has gained much attention in the present context of global changes. Finally, there has been an increasing concern for the conservation biology of cryptogams, and bryophytes in particular.

The aim of the present book is to offer a stepping-stone to anyone interested in discovering the fundamentals of the biology of bryophytes, making the bridge with more comprehensive treaties such as Goffinet and Shaw's *Bryophyte Biology* (2009) and Glime's *Bryophyte Ecology* (2007a). In *Introduction to Bryophytes*, we make an attempt at summarizing and explaining the recent advances made across the various aspects of bryophyte biology at a level that would be palatable to beginners. The focus of the book is thus to open the door to the curious mind and to inspire students, from late undergraduates to post-doctoral researchers, to engage in bryological research.

The book follows what we consider to be an intuitively natural flow. We focus first on the origin of bryophytes in the context of land plant evolution and on the significance of bryophytes in today's world. The morphological diversity of bryophytes are outlined and their associations with fungi and cyanobacteria are summarized, before their ecophysiological features are reviewed. We describe the parameters that shape bryophyte communities and the global distribution of the species that compose them. Finally, we examine the threats facing bryophytes and review approaches designed to conserve them. Some aspects of bryophyte biology merited closer attention and are addressed in box essays. All technical terms are defined in a glossary. Spectacular illustrations of many of the terms used in the present volume can also be found in Malcolm and Malcolm (2006).

Writing this textbook has been a definite challenge and we could not have completed it without Robynn Shannon's diligent editorial comments on substance and form. Her repeated calls for clarification and simplification, along with her improvements of the style, have been essential. We also deeply acknowledge the help of the many colleagues who reviewed different chapters of the book, including Peter Alpert (Physiological ecology), Jeff Bates (Physiological ecology and Bryophytes in a changing world), Richard Beckett (Physiological ecology), Jeff Duckett (Evolutionary significance), Rob Gradstein (Liverworts), Mark Hill (Ecology), Jon Shaw (Biogeography), Line Rochefort (Conservation biology), Hakan Rydin (Ecology), Juan Carlos Villarreal (Hornworts) and Harald Zechmeister (Bryophytes in a changing world). We extend our thanks to Virge Kask for her original drawings and Delphine Aigoin, Neil Bell, Paul Davison, Virginie Hutsemékers, Patrick Degroot, Claude Dopagne, Jan-Peter Frahm, Harald Kürschner, Juul Limpens, Barbara Murray, the Peatland Ecology Research Group, Christine Rieser, Gordon Rothero, Ricardo Rozzi, Martin Simard, Juan Carlos Villarreal and Norman Wickett for sharing some of their photographs and/or providing material for our illustrations.

Our research has been possible over the years with financial support from the Belgian and American National Science Foundations, the Fonds Léopold III, the Belgian Academy of Sciences, the Walloon Ministry of the Environment and the National Geographic Society, for which we are grateful. Finally, we also take this opportunity to thank our families, colleagues and friends who have encouraged, motivated and inspired our research from the beginning of our careers, namely André Sotiaux, Jacques Lambinon, Jonathan Shaw, Dale Vitt and the late Lewis Anderson.

1

Evolutionary significance of bryophytes

Vascular plants, particularly seed plants, dominate vegetation throughout much of the world today, from the lush rainforests of the tropics harbouring a vast diversity of angiosperms, to the boreal forests of coniferous trees, draping the northern latitudes of the globe. This dominance in the landscape is the result of a long evolutionary history of plants conquering land.

Evolution is the result of a suite of incessant attempts to improve fitness and take advantage of opportunities, such as escaping competition and occupying a new habitat. Pilgrims, fleeing the biotic interactions in the aquatic habitat, faced severe abiotic selection forces on land. How many attempts were made to conquer land is not known, but at least one of them led to the successful establishment of a colony. At least one population of one species had acquired a suite of traits that allowed it to complete its life cycle and persist on land. The ancestor to land plants was born. It may have taken another 100 million years for plants to overcome major hurdles, but by the Devonian Period (approximately 400 mya), a diversity of plants adapted to the terrestrial environment and able to absorb water and nutrients, and transport and distribute them throughout their aerial shoots, occupied at least some portions of the land masses. Soon thereafter, plants were freed from the necessity of water for sexual reproduction, by transporting their sperm cells in pollen grains carried by wind or insects to the female sex organs, and seeds protected the newly formed embryo. Angiosperms, with their elaborate flowers and seeds packed with a reserve-filled endosperm, are the last major product of land plant evolution. Their origin dates back to approximately 150 mya, at most. Today, scientists estimate the diversity of flowering plants at about 250–300 000 species, distributed throughout the globe, in virtually all habitats. This evolutionary success story has its roots in the Ordovician Period, with the initial transition to land and the diversification of the earliest land plants.

Bryophytes, which evolved during a pivotal moment in the history of life on earth and have persisted for hundreds of millions years (Renzaglia *et al.* 2007), are considered the closest modern relatives of the ancestors to the earliest terrestrial plants. Furthermore, the ancestor to the vascular plants is thought to have shared many features with bryophytes. These diminutive and often overlooked members of our green world hold the key part in the evolutionary history of land plants: bryophytes mark the transition to land and the origin of vascular plants, and hence, link the seed and vascular plants to their algal ancestors.

In this chapter, we outline the characteristics of bryophytes, contrast them to those of other early land plants and summarize our current understanding of the significance of bryophytes in land plant evolution. The phylogeny of land plants also serves to implicitly reconstruct the transformations of critical morphological or life history features. Several physiological and morphological adaptations provided the fundamental requirements for the colonization of and diversification on land. One such critical innovation may have been the ability to establish close interactions with micro-organisms, such as bacteria and fungi, to ensure access to limited and scattered nutrients. Today, bryophytes are vital at the ecosystem level, from playing an essential role in global biogeochemical cycles by sequestering large quantities of carbon as peat in *Sphagnum* bogs, controlling water and nutrient flow in tropical montane forests where they form luxuriant epiphytic vegetation, to providing shelter for a diversity of micro-organisms (see Chapter 2).

1.1 What do we call a bryophyte?

The term 'bryophyte' has its origin in the Greek language, referring to plants that swell upon hydration (see Section 8.1). 'Bryophytes' is a generic name for plants characterized by a life cycle featuring alternating haploid and diploid generations with a dominant gametophyte (Box 1.1). In fact, bryophytes are the only land plants with a dominant, branched gametophyte, which exhibits a diversity of morphologies unparalleled in tracheophytes (Crum 2001). This feature was long considered indicative of a unique shared ancestry, but the notion of the monophyly of bryophytes has now been strongly challenged. Extant bryophytes belong to either liverworts (Marchantiophyta), mosses (Bryophyta in the strict sense) or hornworts (Anthocerotophyta). These lineages share several characters, some of which have been retained by all other land plants (e.g. an embryo which gives land plants their name 'embryophytes'), and others that are unique (e.g. an unbranched sporophyte, with a single spore producing tissue, or sporangium). As in other extant land plants, the gametophyte lacks stomata. The three major bryophyte lineages

Box 1.1
The bryophyte life cycle

The life cycle of all land plants is characterized by an alternation of generations, in which a haploid phase gives rise, following fertilization, to a diploid phase that eventually undergoes meiosis to regenerate haploid cells. The transition from haploid to diploid generation is characterized by fusion of gametes, yielding a zygote. The reverse transition is marked by meiosis, resulting in spore formation. The gametophyte bears the sex organs and the sporophyte holds the sporangium. In all land plants, sex organs are specialized for producing male or female gametes (the sex cells; Box 1.1 Fig. 1a).

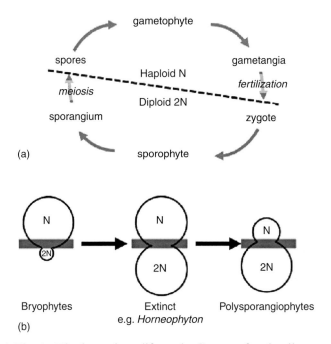

Box 1.1 Fig. 1. The bryophyte life cycle. See text for details.

Bryophytes are the only extant land plants in which the gametophyte is the dominant generation in the life cycle. The sporophyte is unbranched and permanently attached to the maternal (gametophyte) plant. This sporophyte yields a single sporangium. The first multicellular stage following the germination of the spore is termed a sporeling that develops into a gametophyte. In mosses, the sporeling lasts somewhat longer than in liverworts and hornworts and develops into a thallus or a filamentous network termed protonema, which is able to produce several gametophytes. The latter may be branched or unbranched. In all other extant land plants, the balance of dominance is reversed (Box 1.1 Fig. 1b): the sporophyte is branched and independent and the

Box 1.1 (cont.)

gametophyte is reduced. In seed plants, the female gametophyte is enclosed in the ovule and the male cells remain protected by the spore wall (or pollen grain). In some, now extinct, land plants the two generations were very similar, and each was probably independent at maturity.

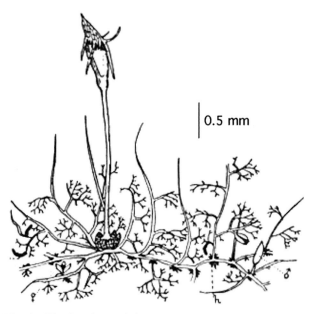

0.5 mm

Box 1.1 Fig. 2. The Southeast Asian moss *Ephemeropsis tijbodensis*, wherein both male and female gametangia as well as sporophytes are directly produced on a persisting protonema. Such a case of neoteny is interpreted as a short-cut in the life cycle, allowing for rapid maturation and reproduction in unstable, temporary habitats such as the twigs and living leaves on which the species grows (reproduced from Goebel 1887).

Because the sperm and egg cells of bryophytes are produced in two distinct organs, and even on two different plants in the case of dioicous species, the sperm must reach the egg for fertilization to occur. Typically, the flagellated sperm cells swim to reach the egg, which must be in close proximity for the fusion to occur. Raindrops crashing onto male sex organs arranged in cups formed by tight whorls of leaves may, upon fragmentation, catapult sperm cells inside tiny droplets over greater distances. Sperm dispersal may also be enhanced by arthropods that are recruited as vectors (Cronberg *et al.* 2006a).

In bryophytes, both generations are usually conspicuous, but reduction of one or the other, or both, is not rare. Loss of complexity characterizes the sporophyte of various lineages of bryophytes, often as an adaptation to dry habitats

Box 1.1 (cont.)

(Vitt 1981). By contrast, some mosses and liverworts exhibit highly reduced gametophytic phases. Some species have the ability to reproduce sexually in the juvenile state. In such neotenic species, the sporophyte develops directly onto the sporeling or protonema (Box 1.1 Fig. 2) (Gradstein & Wilson 2008).

differ from one another in a variety of attributes, most conspicuously in the architecture of the vegetative (gametophyte) body and the sporophyte, to the extent that they are easily distinguished in the field. In essence, the following combinations of morphological and anatomical characters are diagnostic for each of these three lineages:

- **Marchantiophyta** (liverworts; see Chapter 3 and Crandall-Stotler *et al.* 2009 for a more detailed account). The vegetative gametophyte is either thalloid (i.e. ribbon-like plants; Fig. 1.1a) or composed of a leafy stem, with leaves arranged in two or three parallel rows (Fig. 1.1b). Specialized water conducting cells account for endohydric transport in the gametophytes of some taxa (Edwards *et al.* 2003), but are always lacking in the sporophyte. The sporophyte produces a single sporangium elevated, at maturity, on a seta that grows primarily by cell elongation rather than extensive cell divisions. The mode of dehiscence of the sporangium varies but typically, the capsule wall splits along four vertical lines. Stomata are always lacking in the sporangial wall. The capsule holds spores and elaters, elongated cells with spiral wall thickenings that are thought to promote spore release. An axial columella is lacking in the sporangium. Spores typically develop into a single, branched gametophyte.
- **Bryophyta** (in the strictest sense, the division 'Bryophyta' includes only mosses; see Chapter 4 and Goffinet *et al.* 2009 for a more detailed account). The vegetative body is always composed of a stem bearing leaves, typically arranged in spiral rows (Fig. 1.2). Axial water conducting strands occur in both generations of many taxa. The activity of an intercalary meristem, located immediately below the presumptive capsule, gives rise to the seta, which completes its development prior to sporogenesis. This stalk is almost never branched, although in very rare cases two capsules may be found on the same seta (Leitgeb 1876). The sporangium is always terminal. In the majority of mosses, the capsule sheds an operculum (Fig. 1.2). Stomata may occur in the capsule wall, but are always lacking on the seta. The columella typically extends beyond the sporogenous layer. In most taxa, the spore sac surrounds an axial columella, rarely does it also arch over it. Cells of the sporogenous tissue never divide to form elaters. Spore germination results in a filamentous or, in some basal lineages including *Sphagnum* and *Oedipodium*, a thalloid sporeling called a protonema, which subsequently develops into one to several gametophytes.

Fig. 1.1a. The complex thalloid *Marchantia berteroana*, dorsal view, showing hexagonal outlines of the air chambers. Note also the conical gemmae cups (photo R. Rozzi). See plate section for colour version.

Fig. 1.1b. Dorsal view of the leafy liverwort *Leptoscyphus australis* (a), showing lateral leaves organized in two rows. The seta is hyaline and the capsule globose (b) (photo D. Glenny and B. Malcolm). See plate section for colour version.

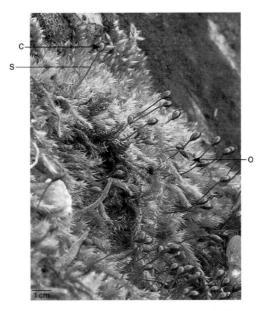

Fig. 1.2. The moss *Sciurohypnum plumosum* with a chlorophyllous seta, S, and capsule, C, closed by a rostrate operculum, O. Note the spiral insertion of the leaves around the stem (photo P. Degroot). See plate section for colour version.

Fig. 1.3. The hornwort *Leiosporoceros dussii*. Note the thalloid gametophyte and the elongate, chlorophyllous sporophyte (photo J.C. Villarreal). See plate section for colour version.

- *Anthocerotophyta* (hornworts; see Chapter 5 and Renzaglia *et al.* 2009 for a more detailed account). The vegetative gametophyte is always thalloid. The thallus may be dissected but never bears leaves (Fig. 1.3). Water conducting cells are lacking in both generations of the life cycle. The sporophyte is linear and composed, except for the foot, of the long sporangium. A seta is thus completely lacking. A basal meristem adds new cells to the base of the sporangium, which consequently matures basipetally. Dehiscence follows two longitudinal lines, extending downward much like a zipper, gradually exposing a spore mass surrounding an axial columella. Pseudo-elaters facilitate the dispersal of spores. Stomata are present on the sporangial walls of some taxa. All hornworts harbour endosymbiotic colonies of *Nostoc*, which form globular or channelled clusters throughout the thallus.

1.2 Bryophytes are embryophytes

The conspicuous morphological disparity between bryophyte and vascular lineages may, at first, obscure their shared evolutionary history. Their common ancestry is revealed, however, by fundamental attributes that these lineages share, as first highlighted by Parenti (1980) and Mishler and Churchill (1984) based on morphology, and subsequently demonstrated repeatedly with DNA sequence data. Land plants clearly share a suite of characteristics, which were all inherited from a single common ancestor, rather than acquired independently by multiple lineages. Morphological features that bryophytes share with other embryophytes (also called land plants) include multicellular sex organs, a cuticle and the retention of the zygote, which undergoes mitotic divisions within the confines of the archegonium. In fact, if the zygote undergoes meiosis (as in most algae), only four spores can be produced. Delaying meiosis, which results in the production of a multicellular body through mitotic divisions, offers the opportunity for more cells to divide meiotically, each meiotic division resulting in four spores. Hence, more spores are produced per zygote, and more spores are dispersed per sexual reproductive event. As more spores are produced, the number of newly established gametophytes is likely to rise. In organisms with unisexual gametophytes, higher population densities may reduce the distance between male and female plants and thereby favour sexual reproduction. Developing a multicellular diploid generation (the sporophyte) with a proliferation of cells undergoing meiosis undoubtedly provides advantages which, over evolutionary time, must have outweighed the cost of resources invested in the production of the sporophyte and spores.

A multicellular sporophyte, however, introduces a weak link in the life cycle: the embryo. Selection on embryo survival must have been so strong that only plants adopting maternal care for the offspring persisted. Indeed, no land plant, whether extant or now extinct, is known to disperse its zygote or its embryo. The zygote develops within the female sex organ. The growing embryo of all land plants is physically attached and physiologically dependent on the female gametophyte (a situation known as matrotrophy), if only during the earliest ontogenetic stages.

Matrotrophy is apparent in bryophytes, as the sporophyte remains physically dependent on the maternal gametophyte from which it receives water, mineral nutrients and various organic compounds. At the junction between the gametophyte and sporophyte, transfer cells forming the placenta enhance the movement of inorganic and organic compounds through the infolded cell wall and membrane, offering a vast surface for exchanges between the two generations (Ligrone *et al.* 1993). A physical connection between the maternal plant and the embryo is required for matrotrophy, hence the zygote, and thus the egg, must remain enclosed in the archegonium. Physical and physiological constraints on matrotrophy may be such that only plants carrying a single egg in their archegonium provide such care. In bryophytes, the maternal plant initiates, upon fertilization, a protective sheath derived from the gametangium and adjacent vegetative tissues around the nascent embryo. The archegonium thus plays a dual role of protecting the egg, and later, the embryo. In liverworts, the sporophyte may complete its development protected by various gametophytic tissues from the archegonium, leaves, or stem (Section 3.1.1). By contrast, the sporophyte of hornworts emerges quickly from the basal sheath, which encloses only the portion of the sporangium undergoing meiosis. In mosses, the modified archegonium or epigonium typically ruptures near its base and the hood (the calyptra) covers the apex of the sporophyte until meiosis is completed. Protection of the young sporophyte seems essential to ensure sporogenesis and thus maximize the chance of producing viable spores. Further evolutionary modifications and improvements eventually led to the ultimate maternal care, as seen in flowering plants. Here, the embryo, along with nutrient reserves and the remainder of the whole female gametophyte, is enclosed within a seed.

Matrotrophy may have been a key innovation acquired early in the evolution of embryophytes, a trait conferring a significant advantage and promoting diversification (Graham & Wilcox 2000). A single lineage with plants nurturing the developing multicellular sporophyte was successful. This lineage gave rise to the land plants as we know them today.

1.3 Bryophytes and land plant evolution

Elucidating the phylogenetic relationships among major lineages of embryophytes, and in particular the relationships among the three main lineages of bryophytes, is essential to reconstructing the origin of critical morphological, anatomical and physiological innovations of plants to conquer land, and thereby to understanding the selection forces that shaped the evolution of plants following the transition to land. Three sources of evidence are available to establish the relationships among early land plants: comparative morphology and anatomy of extant and fossil plants and, more recently, analysis of DNA sequence data.

1.3.1 Bryophytes and the transition to land:
evidence from fossils

Fossilization is an extremely rare event. One just has to look around today to appreciate that few of the individuals living at any particular time will ever be preserved in sediments. The quality of the plant material preserved in geological strata varies, particularly with the mode of fossilization and the time elapsed since the burial. Most Paleozoic Era plant remains occur as compression or impression fossils. Plants were rapidly covered by sediments and, under the growing pressure, were dehydrated and flattened, in some cases reduced to only a thin carbonaceous film, or some cuticular fragments. If all organic material was degraded, a mere imprint of the plant part remains. In aquatic environments, water rich in minerals may have infiltrated the tissues, filled the cells, precipitated and cemented the cell walls. Such permineralized fossils typically retain their three-dimensional architecture, as well as microscopic features of the cell, and therefore offer the most complete picture of early land plant body structure.

Silurian Period (443–416 mya) and younger deposits hold increasingly intact plant remains (megafossils), whereas the Ordovician Period (488–443 mya) and older sediments yield only cellular debris and spores (microfossils; Edwards 2000, Wellman & Gray 2000) (Fig. 1.4). The oldest fragments consist of tubes, tracheids and cuticle. The presence of such decay resistant materials is thought indicative of land plants or their immediate precursors. How these fragments are connected to modern or even extinct lineages that are defined primarily by reproductive features is difficult to establish. Some of the tubular microfossils resemble remains obtained from extant mosses treated with high temperature acid hydrolysis (Kodner & Graham 2001). Similar treatments on several extant bryophytes further revealed a similarity between the fragments resisting the treatment and those recovered from early- and mid-Paleozoic sediments (Graham *et al.* 2004). Spores retaining their tetrahedral arrangement inside an envelope have been recovered from the Ordovician Period (Edwards *et al.* 1995) (Fig. 1.4), and these too find a counterpart among extant liverwort lineages. The microfossil record thus suggests that bryophytes, and in particular liverworts, were integral parts of the earliest plant

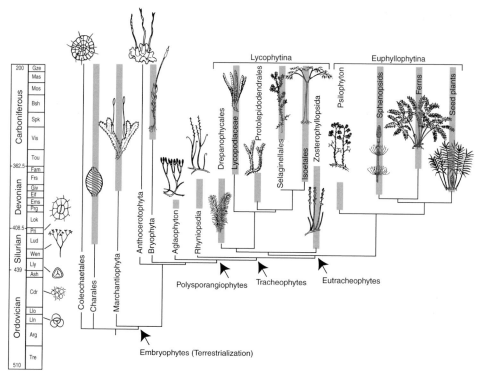

Fig. 1.4. Phylogenetic relationships among major lineages of extinct and extant land plants based on inferences from morphological characters. Grey bars indicate minimum origin of lineage based on megafossil record. Also illustrated along the time scale are minimum age estimates for the appearance of certain important land plant features (from the bottom: spore tetrads, cuticle, trilete mark on spores, megafossils and stomata). Redrawn from Kenrick & Crane (1997b) and adapted based on VanAller Hernick *et al.* (2008).

communities on land. The oldest unambiguous bryophyte fossil is indeed a simple thalloid liverwort (Fig. 1.4). It was, however, recovered from sediments some 30 or more million years younger than those preserving the earliest polysporangiophytes (VanAller Hernick *et al.* 2008). A compression fossil from the Cambrian Period (*c.* 500 mya) has been interpreted by Yang *et al.* (2004) as a possible bryophyte, even a moss, named *Parafunaria* for its resemblance to the extant genus *Funaria*. *Parafunaria* shows whorled leaves and a seta terminated by a capsule. This astonishing discovery pushes the origin of bryophytes, and hence land plants, back significantly, as recently suggested by molecular phylogenies (see Section 1.3.2). It must be noted, though, that the likelihood of fossilization of a particular species depends on its abundance as well as on the presence of decay resistant compounds in its tissues and the suitability of environmental conditions for plant tissue preservation. Hence, the presence of a species is likely to be recorded in the sediment only well after its actual phylogenetic origin. Consequently, fossils provide only a minimum estimate for the time of origin of particular lineages. Reconstructing the transition to land and the early diversification of land plants based on the stratigraphic occurrence of fossils is thus clearly challenged by the fragmentary nature of the material, which obscures the affinities of these fossils to extant lineages of land plants and, in particular, to bryophytes.

1.3.2 Phylogeny of the early land plants

In the absence of an unequivocal fossil record, assessing the significance of bryophytes in the evolution of land plants requires the resolution of the ancestor–descendant relationships among embryophytes. Phylogenetic hypotheses are inferred based on variation in characters scored for the taxa of interest. The nature of the characters sampled may limit the breadth of the taxon sampling and vice versa. DNA sequences offer a potentially vast array of characters, from single nucleotide substitutions to structural changes in the genome in any of the three genomic compartments, the nucleus, chloroplast and mitochondrion. Nucleic acids are generally completely degraded in fossil material, however, and consequently extinct taxa are conspicuously lacking from molecular phylogenies. Similarly, extant taxa may differ in aspects of their development or sperm ultrastructure, two sources of characters inaccessible in fossils. The integration of fossil taxa requires drawing the phylogenetic inference from variation in morphological characters, with the caveat that plant remains may be fragmentary, thereby limiting the pool of characters that can be scored. Furthermore, the gap in morphological complexity between a putative algal ancestor and land plants is such that few characters can be directly compared to determine which lineage of embryophytes marks the earliest cladogenetic event and, hence, the polarity of transformations.

Inferences from morphological characters based on extinct and extant taxa Reconstruction of the evolutionary origin of embryophytes based on morphological characters point to members of the freshwater algae Charophyceae as the putative progenitor lineage to land plants (Parenti 1980, Mishler & Churchill 1984, 1985). The Charophyceae comprise uni- to multicellular organisms, some of which live in soil. All extant species follow a haplobiontic life cycle, wherein the diploid stage is restricted to a single cell, the zygote. Only members of the Charales and Coleochetales retain their zygote, a characteristic that makes them a likely sister taxon to embryophytes (McCourt *et al.* 2004). The Charales also accumulated other morphological, biochemical and physiological innovations,

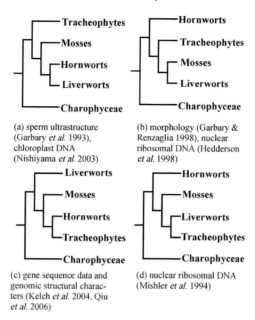

(a) sperm ultrastructure
(Garbary *et al.* 1993),
chloroplast DNA
(Nishiyama *et al.* 2003)

(b) morphology (Garbary &
Renzaglia 1998), nuclear
ribosomal DNA (Hedderson
et al. 1998)

(c) gene sequence data and
genomic structural characters (Kelch *et al.* 2004, Qiu
et al. 2006)

(d) nuclear ribosomal DNA
(Mishler *et al.* 1994)

Fig. 1.5. Summary of alternative phylogenetic relationships among extant lineages of land plants.

such as the ability to develop plasmodesmata or to synthesize phenolic compounds bound to walls. These features may be seen as preadaptations that allowed the initial transition to land (Graham *et al.* 2000, 2006, Ligrone *et al.* 2008).

Mishler and Churchill (1984) and Kenrick and Crane (1997b) proposed that, from a Charophycean ancestor, the three lineages of bryophytes form a grade leading to the vascular plants, with liverworts composing the sister lineage to the remaining embryophytes, and mosses sharing a common ancestor with the tracheophytes (Fig. 1.4). Two transformations supported the monophyly of this combined lineage: a well-developed sporangiophore and the absence of elaters/pseudo-elaters. The relationships of the bryophyte lineages based on morphological data, however, decay when only one or two characters are removed. Subsequent inferences based on characteristics of the development of male gametes led Garbary *et al.* (1993) to resurrect the hypothesis of monophyly of bryophytes (Fig. 1.5a). Broadening the character sampling to ultrastructural and biochemical features, including those of the sporophyte, however, showed hornworts composing the sister lineage to the remaining land plants, with mosses and liverworts derived from a unique ancestor sister to vascular plants (Fig. 1.5b) (Garbary & Renzaglia 1998). These hypotheses collapsed when features of either generation were analyzed separately.

These shortcomings, due to the relatively low number of available morpho-logical and anatomical characters, led researchers to focus on inferring relation-ships based on a much larger source of characters, namely DNA sequences.

Inferences from molecular characters scored for extant taxa In the 1990s, as the molecular revolution swept across the field of systematics, the phylo-genetic signal of nucleotide substitutions was extracted from several loci in attempts to reconstruct the evolutionary history of land plants. The initial euphoria, however, soon dissipated as inferences from distinct loci yielded conflicting and often only weakly supported phylogenetic hypotheses (Goffinet 2000), yet only one of the gene trees could accurately reflect the species tree. Basically, all possible combinations have been proposed (Fig. 1.5; Goffinet 2000). Initial efforts to strengthen the robustness of the tree relied on broadening the character sampling at the expense of taxon sampling, with the result of recovering a monophyletic bryophyte lineage, sister to the tracheophytes (Fig. 1.5a; Nishiyama *et al.* 2003). Inferences from structural changes in the complete chloroplast genome from a similar set of taxa (Kelch *et al.* 2004) led, by contrast, to the hypothesis that hornworts are the closest extant relative of vascular plants. The latest effort, combining an extensive set of DNA sequences from all three genomes, sampled from many representa-tives of the major lineages, provided the most robust hypothesis to date (Qiu *et al.* 2006): hornworts share a common ancestor with vascular plants, whereas liverworts are a sister lineage to all other extant embryophytes. Mosses bridge the gap between liverworts and hornworts (Fig. 1.5c).

Timing the origin of land plants A robust phylogeny of extant land plants not only offers insight into the relative sequence of cladogenic events, but can also be used to date these events using calibration points on the tree. If the substitution rate was known, the time needed for the observed number of mutations to accumulate could be estimated simply by dividing half the genetic distance separating two taxa (or their respective distance to their common ancestor) by the rate of change. Since rates are not known, fossils must be used as calibration points: if seeds appear in the sediments deposited 320 mya, the sum of mutations along the path from the common ancestor to the seed plant to any extant species divided by 320 would yield a rate of substitution per million years. This estimate is conservative given the nature of the fossil. This rate could then be used to infer divergence times of other lineages on the tree. Nucleotide substitutions, however, rarely accumulate at a constant rate in different lineages, so that the so-called 'molecular clock' hypothesis is often violated (Welch & Bromham 2005). Most recent statistical developments take into account heterogeneity in rates of evolution among lineages and, although the conflict among studies highlights the difficulty of reconstructing ancient evolutionary events, offer a surprising result: land plants are much older than suggested by the macrofossil record. Inferences from 27 protein-coding genes in the chloroplast genome (Sanderson 2003) point to an origin of land plants between 425 and 490 mya. This estimate contrasts with that proposed by Heckman *et al.* (2001) and Hedges *et al.* (2004), who hypothesized, based on analyses of 51 nuclear protein-coding genes, that vascular plants had already diverged from mosses at least 200 my earlier (i.e. 707 mya, the 95% confidence interval spanning 515–899 mya), and thus, that the origin of land plants dates to at least 700 mya. The discrepancy may in part lie in the choice of calibration points (plant fossils were used by Sanderson and non-plant divergence times by Heckman *et al.* and Hedges *et al.*). In any case, inferences based on models of evolution of nucleotide sequences push the date for the origin of land plants into the Precambrian, i.e. much earlier than the earliest Silurian megafossil (approximately 420 mya; Kenrick & Crane 1997a) and potential Cambrian fossils (Yang *et al.* 2004).

1.4 Bryophytes and the conquest of land

The convergence towards a single well-supported phylogenetic hypothesis, whereby hornworts share a common ancestor with tracheophytes, mosses form the sister group to this combined lineage, and liverworts mark the transition to land, represents a major step forward in efforts to unravel the diversification of early land plants, and is a prerequisite for reconstructing the evolution of characters. Apart perhaps from the longevity and large size of the persistently chlorophyllous, largely independent sporophytes and the chemical composition of cell walls (Carafa *et al.* 2005), clear structural synapomorphies between tracheophytes and hornworts are lacking. Hornworts differ from all other land plants by the presence of a pyrenoid, a region in the chloroplast where rubisco is aggregated. Pyrenoids are common among algae and their occurrence in hornworts has been seen as evidence for a direct descendance of the Anthocerotophyta from the algal progenitor of embryophytes. The common ancestry of tracheophytes and hornworts, however, suggest that hornworts acquired pyrenoids de novo. The sistership between hornworts and vascular plants thus calls for a re-evaluation of character evolution and their significance in the transition to land.

1.4.1 Descent by common ancestry: the ontogenetic tool kit of land plants

The evolution of land plants is characterized by a rather continuous, even at times apparently dramatic, increase in morphological complexity, culminating in the development of flowers in angiosperms. A comparison of the genome of the moss *Physcomitrella patens* to that of the angiosperm *Arabidopsis thaliana*, reveals that 66% of the approximately 26 000 genes of the latter have a homologue in the moss genome (Nishiyama *et al.* 2003). This genetic similarity that seems incongruent with the morphological gap between these organisms, in fact, merely reflects the shared evolutionary ancestry of their lineages. Part of the genomic overlap is explained by the overlap in cellular metabolic activities. For example, all plant cells must synthesize cellulose to build their new walls following cell division. Hence, it is not surprising that all plants carry the genes coding for these enzymes. Similarly, the physiological mechanisms involved in freezing tolerance are very similar between bryophytes and angiosperms (see Section 8.4.1.2), as are the mechanisms involved in desiccation tolerance in angiosperm seeds and bryophytes (see Section 8.1.3). These mechanisms might have been necessary for the common ancestor of land plants to adapt successfully to non-aquatic conditions and have been subsequently conserved, to a certain extent, in the course of evolution.

Part of the genetic similarity between bryophytes and angiosperms also originates, however, from the genetic connection between even seemingly clearly distinct structures. The root hairs at the apex of roots of the sporophyte of vascular plants appear at best analogous to the rhizoids emerging from the gametophyte of bryophytes. Menand *et al.* (2007) have recently demonstrated, however, that rhizoid development in *Physcomitrella* and root hair initiation and growth in *Arabidopsis* are controlled by the same transcription factors, revealing that gene recruitment may have been an important mechanism by which to assume new functions. In other words, the genes coding for bryophyte rhizoids would have been at the basis of the evolution of root hairs in vascular plants. A more dramatic example is provided by the FLO/LFY genes, which code for transcription factors. In *Physcomitrella*, two LFY homologues control the first mitotic division undergone by the zygote. In flowering plants by contrast, FLO/LFY determine the differentiation of floral parts and, hence, are critical for successful sexual reproduction (Tanahashi *et al.* 2005). This pattern in the evolution of gene function led Floyd and Bowman (2007) to refer to the genes used by all land plants, even if for different functions, as the tools of a fundamental kit needed by plants to develop their vegetative and reproductive bodies.

1.4.2 Independent attempts towards an adapted terrestrial body within land plants

Increasing evidence is thus revealing that bryophyte innovations served, sometimes unexpectedly, as a basis for subsequent evolution in vascular

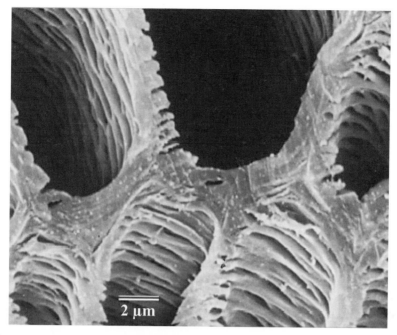

Fig. 1.6. Scanning electron micrograph of water-conducting cells in the simple thalloid liverwort *Symphyogyna*, showing the elongated pits in the thickened walls (reproduced from Ligrone *et al.* 2000 with permission of the authors and The Royal Society, UK).

plants. By contrast, a series of other characters thought to have first appeared in bryophytes and then to have been modified by descent in vascular plants, are now thought to have evolved repeatedly and independently in different lineages. The most striking examples concern the water-conducting structures, stomata and endosymbioses. These characters are, in fact, almost certainly not homologous between bryophytes and tracheophytes and even among the bryophyte lineages themselves. These features are now interpreted as the result of several independent attempts of land plants to adapt to their new, terrestrial condition.

Evolution of conducting structures Bryophytes are often erroneously referred to as non-vascular plants, implying that they are devoid of specialized tissue for the internal transport of water and nutrients. In fact, specialized internal conducting cells are found in some liverworts and most mosses (Ligrone *et al.* 2000). All bryophyte water-conducting cells are dead and lack any cytoplasmic content. Within liverworts, water-conducting cells are present in two main lineages. In some genera of simple thalloid lineages (see Section 3.1.1), water-conducting cells are elongate ($8 \times 300\,\mu m$) with tapering ends and thick walls perforated throughout by numerous pits (Fig. 1.6).

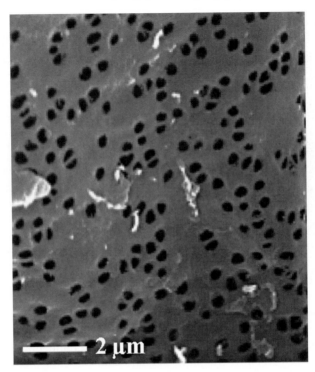

Fig. 1.7. Scanning electron micrograph of the end-wall of a water-conducting cell in the liverwort *Haplomitrium*, showing numerous, large pores (reproduced from Edwards *et al.* 2003 with permission of Blackwell).

In the basal liverwort and moss genera *Haplomitrium* and *Takakia*, these cells are, by contrast, similar in shape to the surrounding parenchymal cells, but are perforated by numerous pits of about 300–600 nm in the former and 120 nm in the latter (Fig. 1.7). Finally, differentiated, highly elongated cells (*c.* 200–1500 × 10–25 μm) called hydroids, with thin, non-perforated walls (Fig. 1.8), are present as a central strand in the stem, seta and leaf costa of most mosses (see Section 4.1).

For many decades, a possible homology between bryophyte water-conducting cells and water-conducting elements of tracheophytes figured prominently in phylogenetic speculation linking these groups. In particular, similarities between hydroids and tracheids have been emphasized to support the contention that these two cell types are homologous, i.e. have the same origin. The assumption of homology of hydroids and tracheids is fundamental to cladograms in which mosses are sister to tracheophytes (Fig. 1.4). There are, however, three main differences between hydroids and water-conducting cells of tracheophytes. First, hydroids lack secondary wall patterns such as spirals, bands or pitting that are characteristic of tracheids. Second, although lignin-like polymers have been detected in bryophytes, they are not tissue

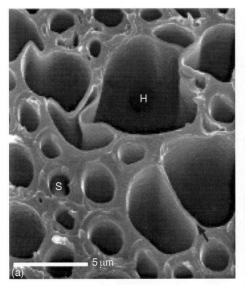

Fig. 1.8. Scanning electron micrograph of a transverse (a) and longitudinal (b) section of a central hydroid, H, strand in the moss *Dawsonia*, showing the completely smooth walls and the thick-walled surrounding stereids, S (reproduced from Edwards *et al.* 2003 with permission of Blackwell).

specific, whereas lignin is predominantly localized in tracheids, vessels and fibres in tracheophytes (Ligrone *et al.* 2008). This suggests that, unlike lignins, the lignin-like polymers found in bryophytes do not fulfil specific structural functions and are more likely to be involved in protection against micro-organisms (see Section 2.1). Third, as opposed to tracheids and vessels, hydroids collapse during water stress and are highly resistant to cavitation, that is the rapid formation and collapse of vapour pockets in water caused by the drop in pressure associated with desiccation (Ligrone *et al.* 2000). This emphasizes a major difference in the water relations between bryophytes and tracheophytes: while the former are mostly desiccation-tolerant, in other words, at equilibrium with external ambient humidity and suspend physio-logical activity upon drying, tracheophytes mostly tend to resist desiccation by pumping water from the soil through roots and limiting water loss by stomata and a waterproof cuticle (see Section 8.1).

Perhaps more striking is the similarity between the water-conducting cells in simple thalloid liverworts and tracheids, which both have thickened walls with helicoidally arranged pits. Such a similarity might suggest that simple thalloid water-conducting cells were the precursors of tracheids, which would definitely lend support to the hypothesis that liverworts are sister to tracheophytes (Fig. 1.5d). The pits of liverwort water-conducting cells, how-ever, develop by removal of secondary wall material closely associated with

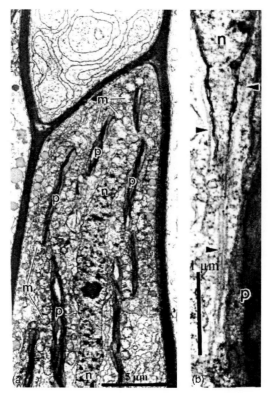

Fig. 1.9. Transmission electron micrographs showing cytoplasmic polarity in food-conducting cells in *Mnium hornum*, with the mitochondrion m, nucleus n and plastids p concentrated at the distal end of the cell (a) and details of the microtubules (arrowheads) extending from the nuclear pole into the cytoplasm (b) (reproduced from Ligrone & Duckett 1994 with permission of *The New Phytologist*).

modified plasmodesmata, whereas in tracheophytes they arise from the lysis of primary unlignified walls with no direct link to plasmodesmata. The two cell types, therefore, have sharply different developmental designs, making homology between them highly unlikely.

Indirect support for the notion of an independent origin for hydroids and tracheids also comes from studies of photosynthate-conducting cells (the sieve cells in tracheophytes) in mosses (Ligrone *et al.* 2000). In mosses, photosynthate-conducting cells exhibit a distinctly asymmetric cytoplasmic organization in longitudinal section, with many organelles and the nucleus close to one end and much less dense cytoplasm at the opposite end (Ligrone & Duckett 1994). This condition establishes two clearly different poles at the end of the longitudinal axis and is therefore referred to as cytoplasmic polarity (Fig. 1.9). From the nuclear pole, microtubules extend as parallel arrays. From these,

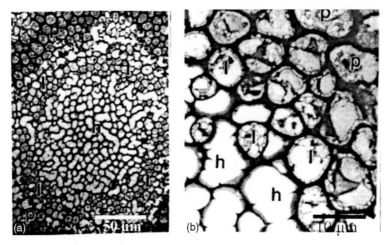

Fig. 1.10. Details of food-conducting cells under light microscopy: transverse section of the stem of *Polytrichum commune* (a) and detail showing leptoids, l, and associated parenchyma cells, p, sheathing around hydroids, h (b) (reproduced from Ligrone & Duckett 1994 with permission of *The New Phytologist*).

single microtubules diverge for a considerable distance in the cytoplasm, sometimes to both ends of the cell. This organization facilitates food conduction, either by movement along stationary microtubules or by microtubule–microtubule sliding. It is best developed in the order Polytrichales (see Section 4.1), wherein highly specialized cells, referred to as leptoids, are intermingled with parenchyma cells and form a wide irregular sheath around the water conducting strands of hydroids (Fig. 1.10) to function in long distance transport of organic nutrients. The sieve cells of tracheophytes, by contrast, are not polarized and microtubules do not persist after maturation of the cell (see Pressel *et al.* 2008b for further details). The two cell types are therefore unlikely to be homologous. The striking similarity between leptoids and sieve cells is probably an instance of homoplasy related to the relatively large sizes attained by Polytrichales and consequent evolutionary pressure for a more efficient transport of photosynthates. A similar example of homoplasy is found in the brown algae, notably the Laminariales, which also contain highly specialized photosynthate-conducting cells bearing striking similarity to sieve elements.

Thus, the hypothesis of multiple evolutionary origins of water-conducting cells in mosses, liverworts and tracheophytes is now strongly supported (Renzaglia *et al.* 2007). Making holes in the walls by disruption of plasmodesmata, together with total loss of cytoplasmic content, is perhaps the easiest way to form a water-conducting cell under selective pressure for more efficient water transport. This suggests that the water-conducting cells that we observe in extant bryophytes correspond to multiple, independent attempts

Fig. 1.11. Stoma in sporophyte epidermis of the hornwort *Leiosporoceros* showing two guard cells surrounding a median pore (reproduced from Renzaglia *et al.* 2009 with permission of Cambridge University Press).

by land plants to adapt to their terrestrial environments and that they cannot be considered as the precursors of the tracheids and vessels of tracheophytes (Ligrone *et al.* 2000). The fact that hornworts, which are the only bryophyte lineage lacking water-conducting cells, are now considered to be sister to tracheophytes, is consistent with such an interpretation.

Evolution of stomata Stomata in bryophytes typically exhibit the same morphology as in tracheophytes: two guard cells surrounding a pore (Fig. 1.11). They are present in the sporophyte of most mosses and hornworts, but are absent from liverworts. Although it is not contested that bryophyte stomata facilitate gas exchange, the homology and function of these structures has been questioned. A major difference between bryophytes and tracheophytes is that the latter tend to resist desiccation by maintaining a high cell water content even when atmospheric humidity is low. In this case, the function of stomata is to maximize CO_2 exchange while minimizing water loss by evaporation and successive opening and closing therefore follows diurnal cycling.

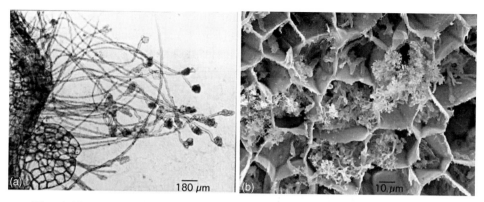

Fig. 1.12. Fungal associations in bryophytes: swollen-tipped rhizoids of the liverwort *Cephalozia connivens* after experimental inoculation with mycorrhizal fungi found on Ericaceae (a) (reproduced from Read *et al.* 2000 with permission of the authors and The Royal Society, UK) and scanning electron photomicrograph of fungus arbuscules in a cross-section of the thallus of the liverwort *Marchantia foliacea* (b) (reproduced from Russell & Bulman 2005 with permission of *The New Phytologist*).

In bryophytes, stomata occur only on the sporangium of some mosses and hornworts and only in parts exposed below the calyptra or above the involucre. They are completely lacking from the vegetative gametophytes. Although the guard cells are able to open and close the stomata (Garner & Paolillo 1973), the pores seem to remain open early in the maturation of the sporangium. This pattern is inherently incompatible with a function solely of gas exchange and water retention. Renzaglia *et al.* (2007) hypothesized that, in bryophytes, stomata favour capsule dehydration to facilitate separation of spores, capsule dehiscence and spore release. The stomata of early polysporangiophytes resembled those of extant vascular plants. Unlike in bryophytes, stomata of these Silurian and Devonian plants were not restricted to the sporangial wall, in fact, they occurred on the gametophyte as well as on the sporophyte (Edwards *et al.* 1998). Stomata may therefore have distinct functions in modern plants. The function of stomata may thus not be homologous among land plants, but the possibility remains that the basic design may be shared and may have been fundamental for further ultrastructural and physiological modifications resulting in the complex stomata of higher plants.

Transition to land and fungal associations Most land plants (90%) establish intimate symbiotic associations with fungi in or around their roots (Wang & Qiu 2006). In bryophytes, various fungi have been reported inside the cells (endomycorrhizae) (Duckett *et al.* 2006a, b, Zhang & Guo 2007). Intracellular fungi have been found in rhizoids, where they induce the formation of swollen apices (Fig. 1.12a). They may also occur in the thallus, where they form ingrowth pegs into surrounding cells that much resemble the peloton structures found in the roots of angiosperms (Fig. 1.12b) (Duckett & Read 1991, Kottke *et al.* 2003). The absence of damage to the host cells and occurrence of the fungus within specific zones indicate a high level of compatibility between the partners (Duckett *et al.* 2006b).

This broad phylogenetic distribution across embryophyte lineages suggests, not surprisingly, an ancient origin of mycorrhizae in the evolutionary history of land plants. Mycorrhizae may, in fact, have been essential for the colonization of (Pirozynski & Malloch 1975) and even the morphological diversification (Brundrett 2002) on land. In a terrestrial environment, where nutrients are much less mobile than in water, the immense surface area of the hyphae increases the absorption or exchange capability necessary for efficient mining of the resources, in particular those macronutrients with low mobility in the soil (Cairney 2000). Land colonization also involved the evolution of a suite of protective mechanisms against pathogens, the development of which might be directly (through the production of antibiotics) or indirectly (by depleting the nutrient pool, or triggering plant defences) prevented by mycorrhizal fungi (Selosse *et al.* 2004).

For mycorrhizae to have played a determinant role in the origin of land plants, two conditions must have been met: (a) fungal partners occurred at the time of the transition to land and (b) the association is exhibited by fossils of the earliest land plants. Phylogenetic inference suggests that the Glomeromycetes, which are the most common endosymbiothic fungi found in bryophytes, originated no later than 500 mya (Berbee & Taylor 2007). Their ancestry is corroborated by fossils from the Ordovician Period (460 mya; Redecker *et al.* 2000). However, the origin of the mycorrhizal lineage may be much younger (Berbee & Taylor 2007), in which case the association of Glomeromycetes with extant lower land plants (Read *et al.* 2000) would result from recent adoption of this strategy rather than from inheritance of a deeply ancestral trait (Selosse 2005). Such an interpretation finds some support in the fact that most mycorrhizal fungi recovered from liverworts belong to a rather derived lineage of *Glomus* comprising species associated with vascular plants (Ligrone *et al.* 2007). The fungus of *Marchantia foliacea*, for example, arose from a group of species that form mycorrhizal associations with flowering plants (Russell & Bulman 2005). Similarly, the fungal symbiont of the hornwort *Anthoceros* is known primarily from associations with higher plants (Schüßler 2000).

By contrast, the fungi recently reported from mosses span a broader phylogenetic spectrum encompassing four (rather than one) families of Glomerales (Zhang & Guo 2007). This perhaps indicates a more ancient origin of the moss–fungal interaction than in the case of hornworts and liverworts. Furthermore, remarkable parallels between fungal associations in the basal liverworts genera *Treubia* and *Haplomitrium* (Section 3.1.4), as well as the very ancient pteridophyte genus *Lycopodium*, suggest that these associations epitomize very early stages in the evolution of glomeromycotean symbioses (Duckett *et al.* 2006a). Finally, fossil evidence suggests that land plants had already established fungal associations by the Devonian Period (Taylor *et al.* 2004, Krings *et al.* 2007) and, at least in one case, the association shared similarities with modern endomycorrhyzae (Remy *et al.* 1994). Whether this ancient interaction was indeed mutualistic can, however, not be determined.

2

Ecological significance of bryophytes

Bryophytes are familiar and attractive components of many natural landscapes such as peatlands, tundra environments and moist forests, where their shaggy coverings on branches, crags and boulders, add distinction to the landscape (Figs. 2.1, 2.2). In particular, epiphytic bryophytes make up a conspicuous portion of many rain forest canopies, hanging from branches (Fig. 8.6) and densely festooning tree trunks (Wolf 1993, Kürschner & Parolly 2004).

Less appealingly, they also occasionally grow on unexpected substrates, like the leather of a discarded boot, or a rusty iron pipe. Even in modern cities where air pollution and the man-made environment may seem unrelenting, bryophytes are able to colonize crevices in masonry (Bates 2000). The occurrence of bryophytes in a wide range of terrestrial and freshwater habitats suggests intriguing questions on how they manage to colonize, adapt and develop in such environments. The occurrence of bryophytes in nearly every ecosystem on earth (all but salt water and permanently frozen ecosystems) also raises questions about their ecological significance in those ecosystems. This is the focus of the present chapter, where we will explore the contribution of bryophytes to ecosystem functioning and, more precisely, how they may influence vegetation dynamics, soil formation and characteristics, global geochemical cycles and hence, climate regulation.

2.1 Water and biogeochemical cycles

Bryophytes contribute to a substantial proportion of the global biomass in a range of ecosystems (Fig. 2.3) and hence play a major role in the cycling of carbon and nutrients through growth and decomposition. Although the contribution of bryophytes to global biomass production varies considerably from one ecosystem to another (Fig. 2.4), moss photosynthesis can represent up to 50% of total ecosystem photosynthesis, for example in swampy black

Fig. 2.1. Spring area on Disko Island (Greenland) covered by the moss *Pohlia wahlenbergii* var. *glacialis* (photo A. Vanderpoorten). See plate section for colour version.

spruce forests (DeLucia *et al.* 2003, Benscoter & Vitt 2007). A large proportion of the CO_2 fixed is stored long-term. The most obvious example of long-term carbon storage by bryophytes is the role of *Sphagnum* in the formation of peat. Peatlands are often characterized by a bryophyte (mostly *Sphagnum* spp.) cover of 90–100% of the ground layer. Peatlands are peculiar ecosystems accumulating organic matter as a consequence of an imbalance between net primary production and decomposition. This is because, although bryophytes lack actual lignin, lignin-like polymers have been identified in the cell walls of mosses, liverworts and hornworts (Ligrone *et al.* 2008). The presence of such compounds, coupled with a high carbon : nutrient ratio, together with unfavourable conditions for decomposition such as low temperature, low pH and anaerobic soil, strongly impede degradation. In acidic peatlands (pH 4),

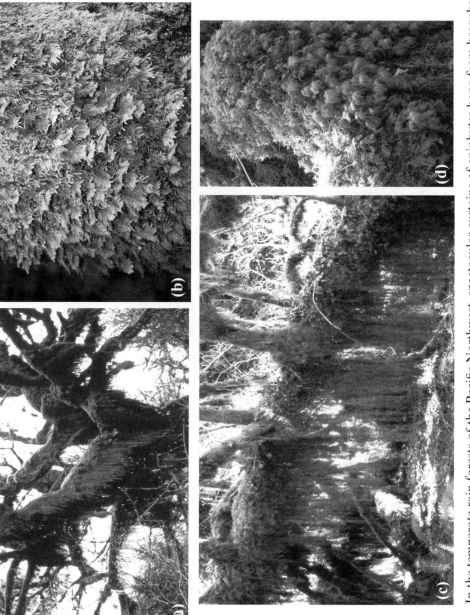

Fig. 2.2. In the temperate rain forests of the Pacific Northwest, spectacular curtains of epiphytes hang from branches and festoon tree trunks (a). (b) Trunk of *Acer* covered by the Pacific Northwest endemic moss *Alsia californica*. (c) Pendent habit of *Isothecium stoloniferum* hanging on branches. (d) Erect habit of *Rhytidiadelphus loreus* and *Hylocomium splendens* hiding the tree base (photos A. Vanderpoorten). See plate section for colour version.

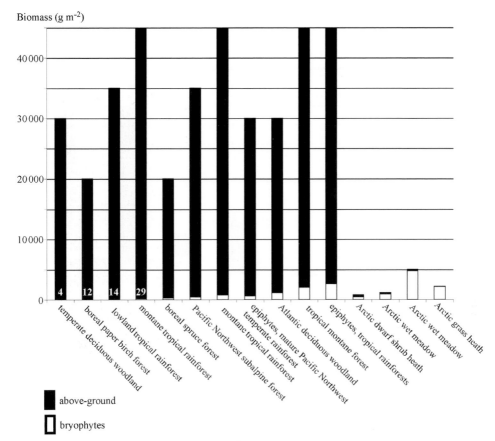

Biomass (g m^{-2})

Fig. 2.3. Bryophyte biomass (g m^{-2}) in a range of terrestrial ecosystems in proportion to above-ground biomass (living + standing dead vascular and lichen vegetation). Data from Longton (1992) for temperate woodlands and Arctic regions; Peck & McCune (1998) for the Pacific Northwest region; Kürschner & Parolly (2004) for tropical rainforests; Rieley *et al.* (1979) for Atlantic woodlands; and Tivy (1993) for global biomasses. Biomass values that are too low to appear on the graph are given within the black bars.

for example, the litter decomposition rate, calculated as percentage of initial litter mass after one year of burial, is only of about 15% in *Sphagnum*, but ranges between 25 and 70% in grasses, forbs and shrubs (Bragazza *et al.* 2007). As a result, *Sphagnum* plays a vital role in global biogeochemical cycles for sequestering large quantities of carbon (Vitt & Wieder 2009). The amount of carbon in northern hemisphere peatlands is 320 Gt, about 44% of the amount held in the atmosphere as CO_2 (Rydin & Jeglum 2006). There is more carbon stored in *Sphagnum* and *Sphagnum* litter than in any other genus of plants, vascular or non-vascular. Over the past millennium, approximately one-third of the world's total carbon has accumulated in the organic deposits

Annual production (g m^{-2})

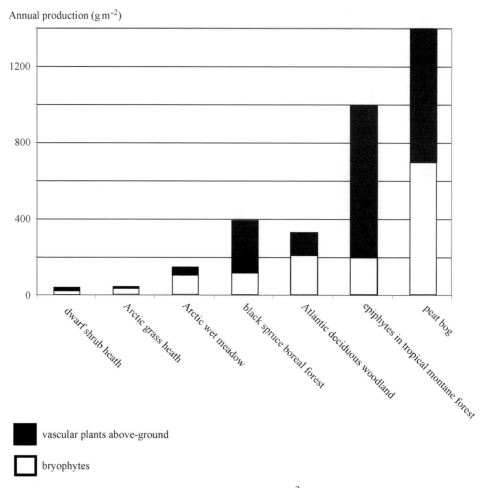

■ vascular plants above-ground

□ bryophytes

Fig. 2.4. Annual bryophyte production (g m^{-2}) in a range of terrestrial ecosystems in proportion to the above-ground production of vascular plants. Data for Atlantic woodlands are for herbs only and those for tropical montane forests for foliage only. Data from Longton (1992) for Arctic regions, Rieley *et al.* (1979) for Atlantic woodlands, Clark *et al.* (1998) for tropical montane forests, and Lavoie *et al.* (2005) for peat bogs.

of peatlands, mainly due to the extremely slow decomposition of *Sphagnum* litter. Since the end of the last glacial period, the soils of the northern latitudes have served as a reservoir for terrestrial carbon, largely through sequestration by *Sphagnum* (Box 2.1). Northern peatlands alone may contain two to three times the amount of carbon stored in tropical rainforests. The conversion of large regions of northern peatlands from net sinks to net sources of CO_2 owing to global warming is, therefore, a major area of concern. On a warming planet, peatlands represent ticking time bombs, as decomposition of the peat

Box 2.1
Economic significance of bryophytes

Bryophytes are of economic interest primarily for their use in fuel production and horticulture. Nearly half the world's annual peat production (Box 2.1 Fig. 1) is used for fuel. In 1995, 71 million m^3 of energy peat were produced, and the carbon flux due to combustion is estimated to be 2.6 million tons per year, equivalent to about 0.4% of the flux due to all fossil energies (Chapman *et al.* 2003). Locally, peat represents a major fuel source and contributed to 25% of the fuel consumption in Ireland in the 1980s (Glime 2007b). Peat has also been increasingly used in horticulture since the beginning of the twentieth century. About 25 million m^3 of peat are used annually as a culture medium, mostly for gardening. Of this, 95% comes from Europe and Canada and the main importing countries include the United States and The Netherlands, with 5.8 and 2.5 million m^3 used annually, respectively (Caron 2001).

Box 2.1 Fig. 1. Peat mining in Quebec (reproduced with permission of the Peatland Ecology Research Group).

Dried bryophytes are also used for cultivating mushrooms, shipping live bulbs and roots, and as a decorative accessory in flora industries. *Sphagnum*, in particular, is an excellent material for the shipment of plants, fresh vegetables

Box 2.1 (cont.)

and flowers, and for hydroponic gardening. Live plants are also used for ornamental purposes in Japanese gardens and aquaria (Gradstein *et al.* 2003) and the cultivation of bryophytes across 1–2 m wide lanes along major roads for trapping dust and micro-particles is currently under scrutiny in Germany (www.moosmatten.de). Bryophytes can be found for sale at nursery suppliers, craft stores and on the internet (e.g. www.hiawathacorp.com/Moss.htm).

At present, although outdoor *Sphagnum* nurseries are an interesting option for a new type of professional horticulture (Rochefort & Lode 2006), bryophytes are directly harvested from the wild rather than cultured. The Pacific Northwest region of North America is, with up to 1470 kg ha^{-1} of harvestable epiphyte biomass in the lower canopy (< 2 m above ground), one of the major areas of bryophytes harvest. Bryophytes were sold in the last decade at prices ranging from USD 0.88 to USD 7.70 per kg. In 1999, the harvest of over 23 million kg (air dried) generated revenues of USD 10 million (Peck 2006).

The common observation that bryophytes are rarely grazed and that they are immune from bacterial and fungal infections owing to the production of chemical compounds with repellent and anti-microbial properties (Parker *et al.* 2000) suggests additional commercial applications. The use of moss and liverwort extracts for protecting crops is one of them, although only a few bryophyte extracts are currently sold commercially (Frahm 2004b). Another application is that, since bryophytes do not require any antibiotics during culture, they can be used to produce targeted molecules without any contamination of the final product. In fact, bryophytes exhibit the highest frequency of homologous recombination among land plants, facilitating targeted gene insertion. Hence, bryophytes can easily be genetically modified in a way that enables them to reproduce exactly the desired protein and are, for example, used to produce antibodies for fighting cancer (www.greenovation.com/english/index.php).

Bryophytes have many additional applications that are too numerous to list here (see Glime 2007b for a more exhaustive account). One of the most recent uses of bryophytes is in forensic science. Genetic fingerprints of bryophyte fragments attached to the shoes or clothes of a suspect can be compared to the DNA sequences of plants found at the crime scene (Virtanen *et al.* 2007).

---

may increase the amount of CO_2 in the atmosphere by as much as 50% (O'Neill 2000).

Bryophytes also play a substantial role in nutrient cycling. In fact, they release nutrients that are made available to other organisms by leaching upon decomposition and, to a more limited extent, by herbivory. In addition, bryophytes sometimes form symbioses with cyanobacteria (Section 3.1.4 and 5.1.4) that are capable of fixing atmospheric nitrogen. The ubiquitous feather moss *Pleurozium schreberi* hosts, for example, *Nostoc* colonies that are responsible for the fixation of one-third to one-half of the $3 \, kg \, N \, ha^{-1}$ annually fixed in boreal forests (De Luca *et al.* 2002).

Even when bryophytes are not dominant, their ecological role in global nutrient cycles can be distinctive and disproportionate to their biomass. In montane tropical forests, for example, the epiphyte bryophyte biomass (a few tons per hectare) is equivalent to 0.5% of the total ecosystem biomass (Fig. 2.3). Obviously, the largest portion of the total forest biomass is the tree component. Yet bryophytes, especially epiphytic species, are of utmost ecological importance regarding the water and nutrient cycles for two main reasons.

First, bryophytes are poikilohydric, which means that they are able to dry out, become 'dormant' and then resume physiological activity upon remoistening (Section 8.1). They display an extremely high water holding capacity that reaches up to about 1500% of their dry weight (Proctor 2009). In epiphyte-rich forests, such as temperate and tropical rainforests, bryophytes may collectively retain about 15 000 kg of water per ha (Kürschner & Parolly 2004, Pypker *et al.* 2006a, b), which is equivalent to the canopy storage capacities in young, closed canopy coniferous forests without significant epiphyte populations. Epiphytes may thus influence the water budget of a watershed by intercepting considerable quantities of rainfall.

Second, the epiphytic bryophyte biomass represents only one to ten years of growth, with a high proportion of minerals in living biomass. By contrast, most tree biomass represents many years of growth and is stored in relatively immobile and unavailable organic forms. Thus, in terms of decomposition rates, epiphytes behave much as the foliar and twig components of trees and understorey members. They are composed mainly of labile materials with relatively rapid turnover times compared to tree branches and trunks. The biomass of all of the epiphytes in montane tropical forests is equivalent to approximately 30% of the foliar biomass (Nadkarni 1984), but the mineral capital contained in the epiphytes constitutes up to 70% of what is held in the foliage of these forests (Fig. 2.5). Thus, epiphytes swell the labile mineral capital of the aboveground portion, expanding the nutrient storage capacity of the ecosystem. Because of the rapid decomposition and heavy leaching pressures characteristic of tropical rainforests, pools of re-utilizable nutrients do not lie in the soil (as in many non-rainforest ecosystems), but in components of the standing vegetation itself. A substantial portion of the mineral capital in many wet forest ecosystems is therefore contained in the living vegetation, with the accompanying potential for long-term adverse effects

Mineral capital (kg ha⁻¹)

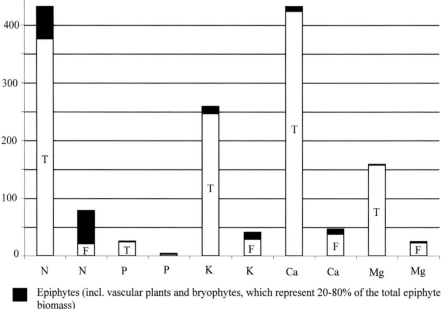

Fig. 2.5. Amounts of mineral capital in epiphytes in proportion to the total above-ground (T) and foliar mineral (F) capital in a Jamaican lower montane forest (data from Nadkarni 1984).

on soil fertility and ecosystem resilience caused by vegetation removal, by logging or land use conversion.

2.2 Vegetation succession and soil formation

Since bryophytes represent a substantial component of several ecosystems, they interact with other plant communities and in many cases have a significant role in soil development. Bryophytes may promote soil formation by accelerating physical and chemical weathering, by trapping wind-blown organic and inorganic material, and by contributing directly to undecomposed organic matter. The pioneer liverwort *Nardia succulenta*, for example, occurs on bare volcanic soils, where it forms layered deposits with volcanic ash up to 15 cm thick. These deposits adhere to vertical cliffs and bridge across volcanic boulders to form a continuous carpet (Fig. 2.6), thus creating a habitat that facilitates the establishment of vascular plants (Jongmans *et al.* 2001).

Fig. 2.6. Bare andesite boulders of a Costa Rican volcano bridged by the pioneer liverwort *Nardia succulenta* (reproduced from Jongmans *et al.* 2001 with permission of Elsevier). See plate section for colour version.

Similarly, bryophytes are of utmost importance in protecting soils in arid areas, where they are components of cryptogamic crusts. These crusts are formed by intimate relationships among a rich assortment of lichens, bryophytes, cyanobacteria, green algae, fungi and the uppermost soil layers. Cryptogamic crusts, which are common in arid and semi-arid landscapes (Fig. 2.7), are the only biological protective cover on the soil surface during unfavourable times such as droughts. They help stabilize the soil surface against wind and water erosion by binding erosion-prone soil particles into more stable soil aggregates. This enhances soil infiltrability and contributes to the redistribution of runoff across the landscape, thereby reducing water erosion (Eldridge 1998).

Bryophytes also play a critical role in mediating both soil temperature and moisture. In fact, bryophytes are effective insulators, promoting the persistence of permafrost by shielding soils from heat during the summer when air temperatures are above freezing (van der Wal & Brooker 2004). Many plant communities of permafrost regions are moss-dominated and the often substantial moss layer can insulate the soil and limit the extent to which it thaws during the summer (Fig. 2.8). Because of the negative impact of cold soil on vascular plant growth, bryophytes might restrict the growth of vascular plants and prevent certain species from establishing in the community. In the most extreme cases, bryophytes are called 'ecosystem engineers' because they

Fig. 2.7. Australian bryophyte soil crusts. (a) General aspect of landscape. (b) Overview of the bryophyte crust (photo D. Eldridge). (c) Photomicrograph (in a plane-polarized light) of a thin section of the bryophyte crust, showing the importance of acrocarpous mosses in the formation of the crust (reproduced from Eldridge 1998 with permission of Elsevier). See plate section for colour version.

are able to create, modify, or maintain certain habitats. This is particularly well illustrated by the active creation and maintenance of wetlands by *Sphagnum* through the process of paludification. Paludification refers to the dynamic process of peat accumulation over a formerly dry mineral soil, which involves the formation of waterlogged conditions (Lavoie *et al.* 2005). A typical example of the result of paludification is the transformation of a black spruce forest into a peatland (Fig. 2.9). Paludification often occurs in wet basins and depressions. In the absence of fire, *Sphagnum* cover increases dramatically and organic matter accumulates at a rate of 0.06–3.8 mm yr^{-1} (Fenton *et al.* 2005). The success of *Sphagnum*, the most important peat-forming plant, is due to its ability to alter its environment to gain a competitive advantage over other plants (van Breemen 1995). Once established, *Sphagnum* species act as ecological switches, modifying the environment to their own advantage through peat accumulation and acidification, resulting in a substantial decrease in species diversity (Fig. 2.10) (Gunnarsson *et al.* 2000). Peat acidification results

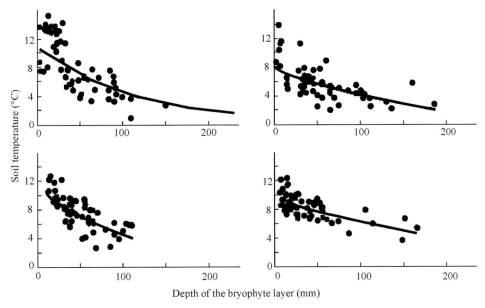

Fig. 2.8. Effects of the depth of the bryophyte layer on the temperature of the top soil layer at four different sites in Spitsbergen (reproduced from van der Wal & Brooker 2004 with permission of Blackwell).

largely from *Sphagnum*'s important cation exchange capacity (see Section 8.3.2), which enables it to take up nutrients and replace them with protons. This can be easily demonstrated by immersing some living or dead *Sphagnum* in distilled water and measuring the pH, then adding common table salt to the same solution and again measuring the pH. In the former case, no pH change is evident, while in the latter, the pH will immediately decrease by 2–3 units (Vitt & Wieder 2009). By reducing soil pH and by promoting cold soil temperatures or even permafrost, *Sphagnum* reduces organic matter decomposition rates, microbial activity and nutrient availability. As the thickness of the organic matter increases, the peat surface is isolated from the mineral soil and the conversion to bog occurs. The accumulating peat may eventually expand beyond the deepest parts of the depression onto the surrounding landscape (Lavoie *et al.* 2005). These changes cause a drop in the growth rate of trees and hence forest productivity (Simard *et al.* 2007), and also inhibit the development of seedlings (Fenton *et al.* 2005). As a result, tree regeneration is hampered and the whole vegetation shifts towards a *Sphagnum*-dominated swamp community.

In some instances, the thick bryophyte cover itself may directly prevent the regeneration of the herb and shrub layers. In relatively dry environments, such as grasslands and heathlands, even a thin moss cover may inhibit

Fig. 2.9. The process of paludification in boreal forests: productive *Picea* forest in Quebec (a) with thick pleurocarpous moss (*Pleurozium schreberi*) cover (b). The accumulation of organic matter progressively leads to water-logged conditions and acidification. As a result, the vegetation shifts towards an unproductive swampy *Picea* forest (c) lying on an accumulating thick peat layer (d) and eventually resulting in a *Sphagnum* swamp (photo M. Simard). See plate section for colour version.

seedling emergence by increasing the time above ground of dispersed seeds, thus increasing the chances of destruction by fire and predation, and by creating a physical barrier to the germination of seeds (Zamfir 2000). In acidic grasslands and heaths, the dense carpets of the invasive moss *Campylopus introflexus* (Section 6.2.2.3) may reduce the germination of heath *Calluna vulgaris* signifi-cantly, up to 60% (Fig. 2.11) (Equihua & Usher 1993). While the impact of the bryophyte layer on the regeneration of the vegetation is undoubtedly negative in such cases, it can also, by contrast, appear as an extremely important factor limiting the invasion of alien species. In Australian grasslands, for example, it was observed that the bryophyte layer has a negative impact on the regeneration

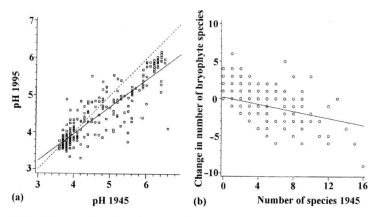

Fig. 2.10. Changes in chemistry and species composition between 1945 and 1995 in a Swedish bog complex. (a) Variation in pH measured at 251 sampling sites. The broken line shows the relationship $x = y$ and the solid line the fitted regression. (b) Relationship between the number of mapped bryophyte species in 1945 and the change in species number by 1995 (reproduced from Gunnarsson *et al.* 2000 with permission of Opulus Press).

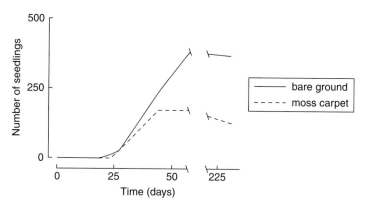

Fig. 2.11. Effect of the invasive moss *Campylopus introflexus* on the germination and survival of seedlings of heath (*Calluna vulgaris*) (reproduced from Equihua & Usher 1993 with permission of Blackwell).

of alien species, but not on native ones (Morgan 2006). It is assumed that the latter have evolved adaptations to allow germination on bryophyte mats through the production of structures such as elaiosomes, which favour burial by ants, or the production of mucus to protect seeds from desiccation. By contrast, alien species lack such adaptations and are therefore restricted to open, disturbed areas.

2.3 Bryophytes as food and as shelter

Anyone who has observed a bryophyte sample under the dissecting micro-scope realizes the extent to which bryophytes serve as a shelter for a vast array of micro-organisms. These range from myxomycetes of tropical rainforests, which display specialized biotypes on epiphyllous liverworts (Schnittler 2001) to invertebrates, including aphids, nematodes, rotifers and tardigrades (Merrifield & Ingham 1998, Peck 2006). In terrestrial environments, the moss habitat is attractive because it provides buffered temperature and humidity conditions. In rivers, aquatic moss cushions reduce water velocity and act as filters of water-borne particles, accumulating large quantities of detritus and periphyton. Bryophytes are also important as both egg-laying sites and a protective nursery for small larvae, which can obtain shelter during floods and consume the abundant food sources available to them. Aquatic bryophytes are attractive refugia for macro-invertebrates because generalist consumers, such as geese and crayfish, selectively consume angiosperms over aquatic mosses, even when the latter represent the bulk of the biomass (Parker *et al.* 2000). The moss tissues, in fact, include a suite of chemical compounds that deter herbivores (Box 2.1). As a result, aquatic bryophytes typically support twice as many plant-associated macroinvertebrates as riverweed (Parker *et al.* 2000) and macroinvertebrate densities are about ten times higher in aquatic bryophytes than on the river bed (Suren 1991).

While bryophytes thus offer appropriate habitat conditions for a variety of invertebrates, direct consumption appears to be remarkably low. Although some invertebrates tolerate or circumvent the chemical defences of bryo-phytes and feed on them (Longton 1992, Parker *et al.* 2000), they most often display a strong dietary preference for angiosperms when confronted with a feeding choice (Smith *et al.* 2001). Snails forced to feed on moss suffer a significant weight loss (Oyesiku & Ogunkolade 2006). Similarly, although accidental ingestion may occur, thereby enhancing dispersal by endozoo-chory (Section 6.2.2.2), vertebrate herbivores have never been shown to graze heavily on mosses, except in Arctic regions. This can result locally in a dominance of bryophytes over grasses in heavily grazed habitats, as shown, for example, in Welsh oakwoods (Rieley *et al.* 1979).

It is neither the mineral content nor the calorific value of bryophytes that make them unsuitable sources of food. Mineral concentrations of bryophytes and angiosperms are, in fact, similar (Table 2.1) and mosses contain the same sugars as higher plants. Lipid levels are reasonably high (5% of dry weight) in the vegetative parts of mosses. In spores, lipid concentrations can reach up to 30% and this may account for the preference for capsules by lemmings.

Table 2.1. *Comparison of the chemical composition of mosses and angiosperms (in % of dry weight) (after Prins 1982)*

	P	Ca	K	Mg	Na	N	crude proteins
bryophytes	0.02–0.15	0.01–0.95	0.36–0.69	0.05–0.22	0.05–0.51	0.96–1.91	5.1–11.9
angiosperms	0.05–0.26	0.14–1.68	1.13–4.16	0.20–0.73	0.03–0.53	1.32–4.1	8.3–28.1

The caloric value of mosses ($3800–4500 \, \mathrm{cal \, g^{-1}}$) is also in the same range as that of angiosperms and no species has been reported to be poisonous. In fact, the main reason bryophytes are not valuable sources of energy is that their digestibility is extremely low (Ihl & Barboza 2007). In reindeer, the maximum net energy gain from mosses is only $78 \, \mathrm{cal \, g^{-1}}$ compared to 1281 cal $\mathrm{g^{-1}}$ in grasses. In lemmings, the difference is $70 \, \mathrm{cal \, g^{-1}}$ in mosses against $552 \, \mathrm{cal \, g^{-1}}$ in forbs. As a result of the poor digestibility of mosses, their consumption may even result in a net loss of nutrients from the animal (Ihl & Barboza 2007).

These differences in digestibility between bryophytes and angiosperms reflect their different strategies for preventing attacks by micro-organisms. Bryophytes mostly lack a cuticle that provides mechanical protection against infection in vascular plants and therefore entirely rely on the use of 'chemical weapons' to prevent microbial and fungal infection. In bryophytes, the high concentrations of polyphenolic lignin-like compounds (*c.* 37% of dry weight in mosses against *c.* 19% of lignin in forbs) provide the cellulose and other polysaccharides with a defence against hydrolytic attack. This also makes the cellulose content less accessible to digestive enzymes. In addition, polyphenols often have an antibiotic action, impeding the digestion of ruminants or hind-gut fermenters. The high affinity of bryophyte cell walls for heavy metals (Section 9.2.2), the presence of other secondary metabolites (Oyesiku & Ogunkolade 2006), and the terpenoid oils contained within the oil bodies of liverworts (Section 3.1.1), may also act as a repellent. Bryophytes therefore appear to be promising 'organic' sources of repellents against snails and other browsing invertebrates (Box 2.1). Only in a few instances have herbivores seemingly adapted specifically to feeding on bryophytes. For example, the relative increase of the rumen weight of Spitsbergen reindeer compared to those of Norway could be seen as an adaptation to a moss diet and the post-caecal spiral in bryophage lemmings may help in the digestion of mosses (Longton 1992).

Quite surprisingly, bryophytes are freely consumed in quantity by Arctic and alpine vertebrate herbivores such as reindeer, caribou, musk ox, most Arctic-breeding geese and lemmings. During the winter, mosses such as

Calliergon, Dicranum and *Polytrichum* form 30–40% of the diet of the latter (Longton 1992). Why then, if Arctic and alpine herbivores do not feed on bryophytes for energy, do they ingest such large quantities of bryophytes? The presence of a specific chemical compound, arachidonic acid, is a possible reason why these plants are eaten under cold conditions (Prins 1982). Arachidonic acid is also present in algae, ferns and *Ginkgo biloba*, but is absent from other gymnosperms and all angiosperms. Among the plants in which it has been reported, the concentration of arachidonic acid is highest in mosses (up to 35% of the fatty acids). Arachidonic acid has interesting properties for life under cold conditions. The low melting point of the molecule ($-49.5°C$) might contribute to the lowering of the melting point of fats, which aids in maintaining sufficient limb mobility at low temperature. Arachidonic acid also offers protection to cell membranes against cold, enhancing membrane fluidity and thus permitting continued enzyme function at low temperatures. Since animals are not capable of synthesizing arachidonic acid, it is adaptive for those living permanently in cold regions to eat moss.

3

Liverworts

The common ancestor to the Marchantiophyta arose immediately after the conquest of land, some 450 mya. Their evolutionary history is, however, also characterized by a more recent diversification during the early Tertiary Period. This diversification may coincide with the establishment of modern rainforests and resulted in species rich crown groups such as the Lejeuneaceae (Heinrichs *et al.* 2007). With approximately 5000 extant species currently recognized, liverworts compose a diverse lineage of land plants, represented on every continent and in nearly all ecosystems. Some species are virtually invisible to the naked eye, whereas others rival mosses in their stature. Although the experienced eye may easily distinguish liverworts from other bryophytes in the wild, not a single conspicuous unifying morphological vegetative character defines them. The architecture of the gametophyte varies from thallose, with or without internal differentiation, to leafy, with leaves divided in some lineages into two or more lobes. It is only after sexual reproduction is completed that liverworts reveal their shared ancestry: the sporangium is enclosed by an epidermis lacking stomata and elevated above the maternal protective tissue on a translucent seta that elongates after sporogenesis. Furthermore, the capsule lacks a columella and the sterile cells within the sporangium develop into elaters. Except for the Blasiales, liverworts lack endophytic cyanobacteria, but most species establish symbiotic associations with mycorrhizal fungi. Although inconspicuous at the landscape level, liverworts form diverse assemblages in moist habitats such as temperate and tropical rainforests, where they thrive on trunks and leaves.

3.1 Structure and development

3.1.1 Gametophyte

The long evolutionary history of liverworts is characterized by conspicuous innovations and subsequent modifications in the architecture of the gametophyte, which have resulted in a striking diversity that is unmatched among land plants, even among the more speciose mosses and the phylogenetically more derived hornworts. The vegetative body of liverworts is either flattened into a thallus with little or no elaborate internal differentiation, or terete and lined with foliar appendages (Fig. 3.1). Growth occurs through the mitotic activity of a single apical cell. The shape of the latter varies and so does the number and orientation of the cutting faces that determine the arrangement of daughter cells contributing to the plant body. The geometry of the apical cell is conserved within the leafy (tetrahedral), complex thalloids (cuneate) and one lineage of simple thalloids (Metzgeriopsida; lenticular), but variable within another lineage of simple thalloids, namely the Pelliidae (tetrahedral, cuneate, or hemidiscoid) (Fig. 3.2).

The gamete-bearing plants of most liverworts are branched either dichotomously, sympodially through subapical innovations, or monopodially with multiple ramifications occurring along the stems. In some cases, branches are dimorphic, with ventral ones acting as anchoring organs (Fig. 3.1f).

Anatomically, the vegetative axis, or thallus, of liverworts is lined by a unistratose epidermis, but otherwise lacks much differentiation, except, as the name indicates, in complex thalloids. In *Marchantia* and its relatives, the thallus is composed of a ventral layer of hyaline storage cells, a dorsal layer of green cells and a unistratose epidermis. The photosynthetic layer includes air chambers that open through superficial pores. The reticulate pattern on the dorsal surface of the thallus (Fig. 3.1c) marks the distribution of the air chambers below. Simple thalloids lack such conspicuous differentiation. At most, a multistratose midrib separates the two unistratose wings (Fig. 3.1d). In some taxa, strands of elongated water-conducting cells, which are absent from complex thalloids and leafy liverworts (except *Haplomitrium*), are formed (Section 1.4.2). In others, the thickness of the thallus gradually tapers towards the margins.

Leafy liverworts compose the largest group, including about 80% of the taxic diversity (Gradstein 1990). Unlike mosses, the leaves develop along three longitudinal, rather than spiral, parallel lines and the stem bears two lateral and one ventral rows of leaves. The latter, termed underleaves or amphigastria, are either similar (isophylly, e.g. *Haplomitrium*; Fig. 3.1a) or different (anisophylly, e.g. *Bazzania*, *Lejeunea* and *Porella*, Fig. 3.1f, g, i) in size and shape from the lateral leaves. Underleaves are lacking in many taxa. The insertion of the underleaves is always transverse, whereas the lateral leaves are commonly inserted obliquely with the line forming either an acute or obtuse dorsal angle with the axis depending on the orientation of the apical cell. In the former case, known as incubous insertion, the forward or upper

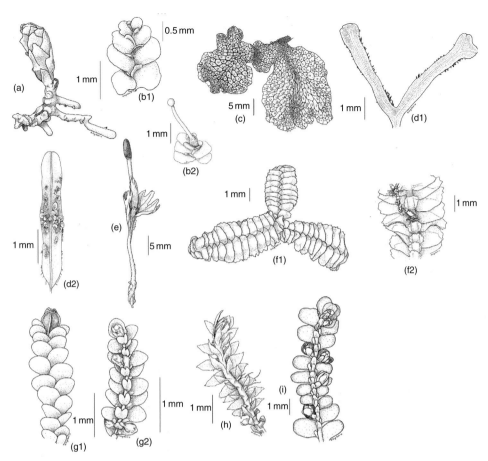

Fig. 3.1. Gametophytic diversity of liverworts. (a) *Haplomitrium gibbsiae*, terete foliose gametophyte with stoloniferous axes and leaves and under-leaves of the same shape and size (isophylly). (b) *Treubia lacunosa*, dorsi-ventrally flattened plant with leaves in two rows divided with a small dorsal lobe (b1) and subterminal sporophyte (anacrogyny) (b2). (c) *Conocephalum conicum*, a complex thalloid, showing the reticulations and pores of the air chambers on the dorsal surface. (d1) *Metzgeria furcata*, with a thalloid gametophyte lacking internal differentiation, except for the midrib. (d2) *Metzgeria crassipilis* bearing numerous discoid gemmae on the thallus surface. (e) *Jenssenia wallisii*, a simple thalloid species growing upright from a terete base. (f) *Bazzania trilobata*, dorsal view (f1), showing indented leaves and ventral view (f2), showing underleaves that are toothed at the margin. Note also the presence of a stolon, ventral branch acting as an anchoring organ. (g) *Lejeunea cavifolia*, dorsal view, plant with perianth (g1) and ventral view showing large dorsal leaf lobes, small ventral lobules and bifid underleaves with entire margins (g2). (h) *Douinia ovata*, dorsal view, with a dorsal leaf lobe that is perpendicular to the larger ventral lobe. (i) *Porella platyphylla*, ventral view, showing small ventral lobules of lateral leaves and fairly large, entire underleaves.

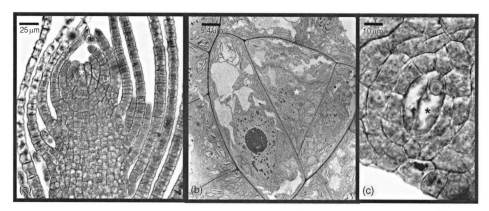

Fig. 3.2. Tetrahedral (a, b) and lenticular (c) apical cells (marked with a*) in the leafy liverworts *Porella* and *Haplomitrium* and in the simple thalloid liverwort *Pallavicinia*, respectively (reproduced from Crandall-Stotler *et al.* 2009 with permission of Cambridge University Press).

edge of each leaf overlaps the rear or lower margin of the next highest leaf along the stem. As a result, the upper edge of each leaf is visible from above, whereas the lower edge is obscured by its older neighbouring leaf (Fig. 3.3b). By contrast, the front margin of a succubous leaf is hidden by the rear margin of the younger leaf. Hence, only the rear margin of each leaf is visible from above (Fig. 3.3a). Incubous and succubous leaf positions are associated with different orientations of the insertion lines of the leaves (Fig. 3.3). In rare cases, leaves are highly reduced, occurring as a pair of cells in *Zoopsis*, which consequently appears almost thalloid (Schuster 1984a). The leaf lamina is almost always unistratose, composed of isodiametric cells. A midrib is lacking, although a vitta of multiple rows of elongated cells occurs in some species. Cell shape varies little across the lamina. Dimorphism is generally restricted to the marginal cells forming a more or less conspicuous border, or projecting outward into teeth or more rarely cilia. The laminal surface is smooth, except in some taxa with mammillose or papillose cells. The cell walls are often thickened in the corners and such trigones occasionally bulge inward.

Because leaves in leafy liverworts grow from two to three leaf-initial cells (only one leaf-initial cell in mosses), leaf morphology is highly variable. Lobed or finely segmented leaves, which are unknown in mosses, are common among liverworts (Fig. 3.4). The division of the leaf lamina into two or more lobes triggered a vast array of modifications of leaf form, including functional divergence between the lobes. Unequal leaf lobing characterizes many taxa. In most lineages, the dorsal lobe is larger (e.g. *Porella*, Fig. 3.1f), whereas in others it is smaller than the ventral one, such as in *Scapania* or *Douinia* (Fig. 3.1h). In two independent lineages of leafy liverworts (Heinrichs *et al.*

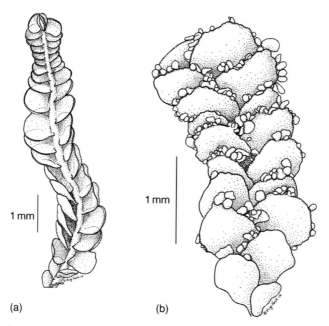

Fig. 3.3. Orientation of leaf insertion. (a) Succubous in *Odontoschisma denun-datum*. (b) Incubous in *Radula complanata*. Note the presence of numerous gemmae at the leaf margin.

2005, He-Nygrén *et al.* 2006), one or two lobes of the divided leaf have been modified into a structure that has long been interpreted as a water-sac (Goebel 1887). Since bryophytes are poikilohydric, this would allow the plant to remain physiologically active for a longer time. Although evidence is lacking to demonstrate the function of these sacs, it is interesting to note that these modifications may disappear in wet environments. In the genus *Frullania*, for example, the ventral lobule is usually helmet-like, but completely unfolded lobules, described as explanate, are produced on plants thriving in moist habitats, such as foggy evergreen forests (Fig. 3.5). In *Colura*, these sacs are further modified. A lid attached by hinge-like cells opens upon hydration and closes as water evaporates, trapping protozoans (Fig. 3.6). Whether this can be truly interpreted as a case of carnivory, a feature as yet only reported in angiosperms (Barthlott *et al.* 2000), remains, however, unknown. The lack of conclusive physiological evidence calls for further research, including a demonstration of the ability of the liverwort to produce proteases, or its ability to assimilate the nutrients released after the degradation of protozoans.

Underlying this architectural diversity of the vegetative body are two microscopic features shared by nearly all liverworts: unicellular rhizoids and oil-bodies. Rhizoids are lacking only in *Haplomitrium*, although they may be

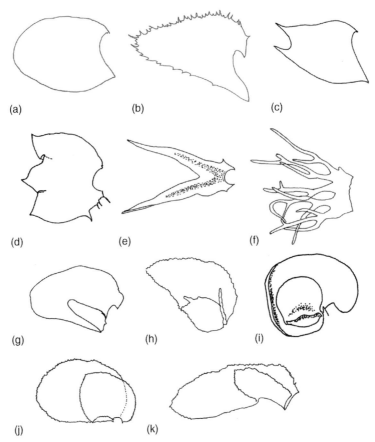

Fig. 3.4. Diversity of leaf shape in liverworts. (a) Undivided, oblong leaf with entire margins in *Jamesoniella autumnalis*. (b) Undivided but toothed leaf in *Plagiochila sharpii*. (c) Lobed leaf with broad arched sinus in *Lophocolea cuspidata*. (d) Leaf divided into three short triangular lobes in *Barbilophozia floerkei*. (e) Deeply bilobed leaves with a forked vitta of elongated cells in *Herbertus aduncus*. (f) Deeply divided leaf with four uniseriate lobes, each further divided into cilia also one cell wide in *Blepharostoma trichophyllum*. (g) Lobed leaf with a small, elongated ventral lobe in *Porella swartziana*. (h) Lobed leaf with a sac-shaped ventral lobule and uniseriate style in *Cololejeunea biddlecomiae*. (i) Lobed leaf with ventral helmet-shaped lobule and short style in *Frullania appalachiana*. (j, k) Lobed leaf with small dorsal lobe and toothed lobe margins in *Scapania nemorosa* (j) and *Diplophyllum obtusatum* (k) (redrawn from Schuster 1966–1992).

lost in some aquatic forms of *Riccia*. Pluricellular rhizoids are known only from *Plagiochila* and *Pachyschistochila* (Schuster 1984a), but only following fungal infection (Pressel *et al.* 2008a). Rhizoids arise from cauline epidermal cells or from the specialized cells typically confined to the base of underleaves; they occur in fascicles or are scattered. The rhizoids of most complex thalloid liverworts are dimorphic, the typical form having smooth walls and the

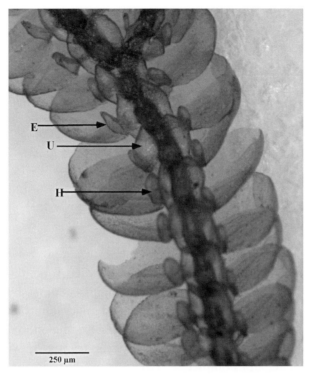

Fig. 3.5. Helmet-like (H) and explanate (E) ventral lobules on the same shoot of *Frullania polysticta*, an endemic species of the foggy laurel forests of Madeira and the Canary Islands. Note the large underleaves (U) that are bifid at the apex (photo A. Vanderpoorten).

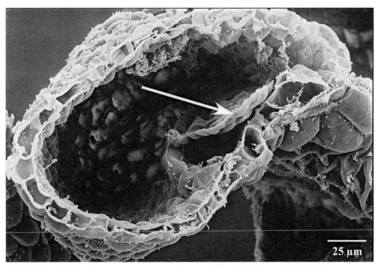

Fig. 3.6. SEM photograph of a transverse leaf section in the liverwort *Colura zoophaga*. The arrow points to the movable lid closing the opening of the sac (reproduced from Barthlott *et al.* 2000 with permission of G. Thieme Verlag).

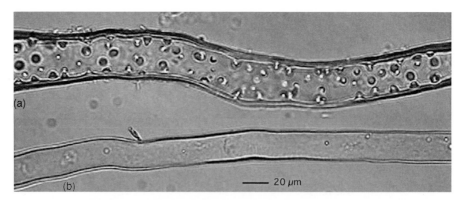

Fig. 3.7. Pegged (a) and smooth (b) rhizoids in complex thalloid liverworts (photo P. Davison).

so-called pegged rhizoids bearing intracellular wall projections (Fig. 3.7). As in other bryophytes, rhizoids likely serve a dual function in liverworts, anchoring the gametophyte to the substratum, sometimes by means of an expanded apical disk, and conducting water and mineral nutrients. In complex thalloids, pegged rhizoids recover their structural and functional integrity after dehydration and seem to be involved in water transport, whereas smooth rhizoids tend to remain completely flattened and anchor the thalli to their substrata (Duckett *et al.* 2000). The smooth rhizoids also serve as gateways for endophytic fungi (Russell & Bulman 2005; see below). In contrast to mosses and hornworts, liverworts almost never develop tubers.

Whereas unicellular rhizoids are a plesiomorphic feature shared with hornworts, oil bodies are truly a unique feature of liverworts among land plants. In contrast to the lipid bodies of other embryophytes, oil bodies are true membrane-bound organelles. They contain terpenoid oils suspended in a carbohydrate and/or protein-rich matrix. Ninety per cent of liverworts develop oil bodies. Their size, shape, number, colour, distribution and chemical composition vary among taxa (Fig. 3.8) and therefore these characters offer critical taxonomic information. For instance, complex thalloid liverworts exhibit peculiar oil bodies, which are found only in scattered idioblastic cells (i.e. cells that differ markedly from neighbouring ones). These cells lack plastids and are almost filled by the single, large, granular to globular brownish oil body (Fig. 3.8e). Similar oil bodies occur in some leafy liverworts, forming so-called 'ocelli' in the leaf. In contrast to the oil bodies of leafy liverworts, however, the matrix of those of complex thalloid liverworts lacks proteins, but contains polysaccharides and polyphenols. Unfortunately, because of the volatility of the oils they contain, oil bodies progressively disappear upon drying in the laboratory. In some taxa, the process takes only

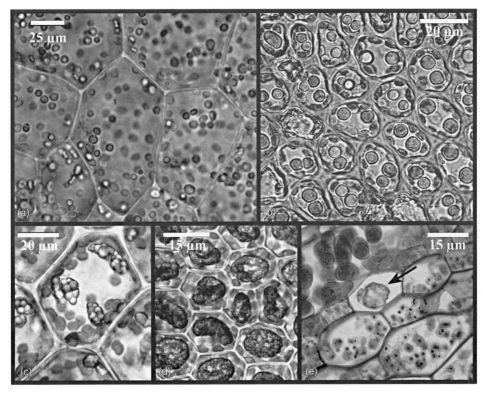

Fig. 3.8. Range of variation of liverwort oil bodies. (a) Numerous, small and homogeneous oil bodies in *Austrofossombronia*. (b) Numerous, large oil bodies in *Marsupella*. (c) Grape-shaped oil bodies in *Calypogeia*. (d) Large, solitary, papillose oil bodies in *Jungermannia*. (e) Large, solitary oil body in an idioblastic cell (arrow) in *Marchantia* (reproduced from Crandall-Stotler *et al.* 2009 with permission of Cambridge University Press).

a few hours, so that fresh material must be studied, whereas in others, oil bodies last for some years and can still be studied in herbarium specimens. The sometimes rapid disappearance of oil bodies in herbarium specimens has long puzzled taxonomists since these organelles invariably remain present in wild material, even after long periods of drought. A series of desiccation experiments most recently showed, in fact, that oil bodies remain intact as long as specimens become desiccated under natural conditions (Duckett *et al.* 2006c). The function of oil bodies remains largely unknown. They are thought to deter herbivores or provide protection from cold and/or UV radiation.

Leaves and rhizoids are not the sole appendages developed by liverworts. Complex thalloid liverworts bear ventral scales, often arranged in two or more rows. These unistratose structures may create capillary spaces essential for external water conduction or prolonged water retention, or may protect the apices (Schuster 1984a). In addition to the volatile terpenoids held in oil bodies, liverworts also produce abundant mucilage, thought to aid water retention. In complex thalloid liverworts, the mucilage is secreted internally by scattered slime cells. In other liverworts, these cells develop from epidermal cells at the apex or close to sex organs. These slime papillae are either stalked or sessile. Here, the ability of the mucilage to retain water serves to protect regions of active cell division from dehydration. Finally, some species develop hairs or leaf-like appendages on their stem or thallus or on the tubular outgrowth protecting the developing embryo. Intuitively, these seem to favour external water conduction, but their green colour suggests that they also contribute to carbon fixation.

As is characteristic of all seedless embryophytes, egg and sperm cells develop within multicellular organs. These gametangia are formed from superficial cells and develop on the thallus surface. In thalloid liverworts, they are often clustered in chambers. About two-thirds of liverworts are dioicous: male and female sex organs occur on distinct individuals. Bisexual plants typically carry male and female gametangia on different branches rather than intermixed in a single cluster. Paraphyses or modified slime hairs are rarely mixed among the gametangia. Antheridia consist of a stalked and more or less spherical chamber, which mostly lacks a differentiated zone of dehiscence. Upon hydration, mature antheridia rupture at their apex and release the sperm cells. In *Conocephalum*, the latter may be projected 15 cm upward and become airborne, greatly increasing the chances of reaching a female plant (Shimamura *et al.* 2008). In *Haplomitrium*, antheridial bracts form rudimentary splash-cups onto which raindrops crash and, upon fragmentation, catapult sperm cells inside tiny droplets over great distances. In some complex thalloid liverworts, the antheridia are concentrated in chambers on the upper surface of receptacles, which in some species are elevated on a stalk. These stalks likely favour the dispersal of sperm to neighbouring female plants. In simple thalloid liverworts, the antheridia are either protected by scales on the thallus surface or enclosed in ostiolate chambers. In leafy liverworts, antheridia are located in the axil of specialized leaves, the male bracts, on the main shoot or on specialized male branches (Fig. 3.9).

Archegonia consist of a venter that is stalked and mounted by a narrow neck. At maturity, the central canal cells disintegrate and the cover cells are released. The distribution of the archegonia parallels that of the male sex organs. In simple thalloid liverworts, they develop on the dorsal surface. In complex thalloid liverworts, archegonia are initiated dorsally and then often displaced to the ventral surface of a specialized, sessile or stalked receptacle, the archegoniophore. The stalk grows after fertilization, except in *Marchantia*

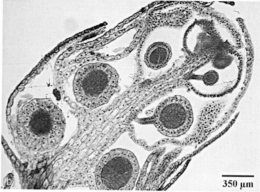

Fig. 3.9. Antheridia in leafy liverworts. (a) Male plant of *Jungermannia* sp., showing an antheridia (arrow) located at the base of a concave male bract (photo P. Davison). (b) Longitudinal section of the shoot, showing stalked antheridia (photo P. Davison).

and *Wiesnerella*, wherein elongation of the archegoniophore is initiated prior to sexual maturity. The development of archegonia in leafy liverworts is initiated by the apical cell of either the stem or the branch, a condition referred to as acrogyny. By contrast, anacrogyny implies the derivation of archegonia from lateral cells. Consequently, acrogynous taxa (all leafy liverworts and a few Metzgeriidae) develop a single cluster of female sex organs at the apex of each axis, compared to multiple lateral groups in anacrogynous species, such as the Pelliidae. The gametangial clusters are typically protected by modified lateral and underleaves, the bracts and bracteoles, respectively, or by leaf-like outgrowths of the thallus. Fusion of leaves internal to the bracts through the coalescence of their primordia produces a perianth, which is typically short at first but resumes growth following fertilization, providing protection for the developing sporophyte (Figs. 3.10, 3.11a). Beyond their protective role, bracts and perianth may also facilitate fertilization by creating capillary spaces converging at the archegonia (Schuster 1984a). Perianth shape and ornamentation vary substantially among liverworts and provide critical taxonomic characters (Fig. 3.10).

The sporangium matures within the confines of a membrane derived largely from the archegonial venter, the calyptra (Fig. 3.11a). The young sporophyte of various leafy and some simple thalloid liverworts may also be partially sheathed by a coelocaule, also called a perigynium, which results from mitotic activity of stem tissue below the cluster of archegonia, ultimately embedding the sporophyte (Fig. 3.11b). In some cases, this cauline growth merges with the archegonial tissue enclosing the egg and forming a thick, fleshy shoot calyptra, which differs, as the name indicates from a true calyptra

(a)   (b)   (c)

Fig. 3.10. Variation of perianth shape in liverworts. (a) Smooth, laterally compressed perianth of *Leptoscyphus australis* with a laciniate mouth margin (photo B. Malcolm). (b) Plicate perianth (arrow) of *Dinckleria* (reproduced from Crandall-Stotler *et al.* 2009 with permission of Cambridge University Press). (c) Ventral side of perianth in *Frullania polysticta*, showing two lateral and a ventral deep keels (photo A. Vanderpoorten).

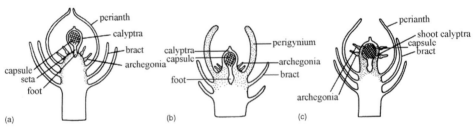

Fig. 3.11. Gametophytic tissues protecting the young sporophyte in leafy liverworts. (a) Perianth. (b) Perigynium. (c) Shoot calyptra (reproduced from Gradstein *et al.* 2001 with permission of the New York Botanical Garden and the authors).

by the presence of stem tissue (Fig. 3.11c). In thalloid liverworts, the archegonia are often surrounded by a tubular or scale-like structure called the involucre. In some genera, there are two involucres. The inner one is called the pseudo-perianth, and develops only after fertilization (Fig. 3.12). It resembles the perianth of the leafy liverworts, but the latter originates from fusion of leaves, whereas the pseudoperianth originates from the thallus (Gradstein *et al.* 2001). All these protective gametophytic structures have arisen multiple times during the diversification of liverworts. They have no equivalent among other bryophytes, and are in fact unique among land plants.

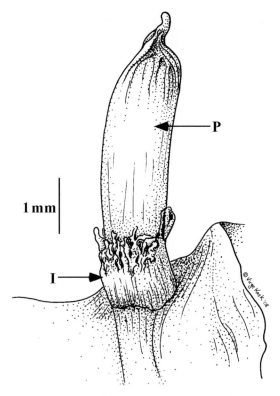

Fig. 3.12. Protective gametophytic structures around the archegonia and, later, the young sporophyte, in the simple thalloid liverwort *Pallavicinia lyellii*, comprising an involucre, I, and a pseudoperianth, P.

3.1.2 Sporophyte

In contrast to the architectural diversity of the vegetative gametophyte, the organization of the sporophyte into a foot, an unbranched seta and a capsule, is highly conserved. Foot and seta are lacking only in the Ricciales, wherein the capsule develops within the thallus. The first division of the zygote is always transverse and yields a hypobasal and an epibasal cell. Subsequent growth never involves an apical cell or an intercalary meristem as in other bryophytes. The significance of the two intial cells in the development of the foot, seta and sporangium varies among lineages, with the hypobasal cell typically forming the foot and seta in complex thalloid liverworts, but solely a haustorium consisting of elongated cells around the foot in *Haplomitrium* and various simple thalloid liverworts (Schuster 1984a). The foot anchors the sporophyte onto the maternal plant. Absorption of nutrients and water may be enhanced by modification of the placental cells marking the junction of the two generations, in particular by wall ingrowths on one or both sides, which

Fig. 3.13. The complex thalloid *Conocephalum conicum*, dorsal view, with a conical carpocephalum. A tubular involucre (arrow) encloses the nearly mature sporophyte (reproduced from Crandall-Stotler *et al.* 2009 with permission of Cambridge University Press).

increase the surface area for exchange (Frey *et al.* 2001). However, actual penetration of gametophytic tissue by sporophytic cells, as occurs in hornworts, is lacking.

Cells of the seta are entirely thin-walled parenchyma, lacking any differentiation. The seta elevates the capsule above the protective gametophytic tissues only after spore maturation is complete through simple but spectacular elongation rather than division of the cells. Through increased turgor due to water intake, the cells of the seta may elongate within a few hours by a factor of 20. In complex thalloid liverworts, which elevate their sporophyte high above the thallus via the archegoniophore (Fig. 3.13), substantial elongation of the seta is not necessary and consequently the seta remains very short. Similarly in the Porellales the seta is generally very short as an adaptation to the epiphytic habitat of the majority of the members of this order and merely ensures that the capsule dehisces beyond the protective tissues (Fig. 3.14).

The capsule (Figs. 3.10a, 3.14) consists of a wall derived from the amphithecium and a sporangium of endothecial origin. A columella is absent. Each cell of the sporogenous tissue divides to yield a sporocyte (spore mother-cell) and an elaterocyte (elater mother-cell) (Figs. 3.14, 3.15). Each elaterocyte

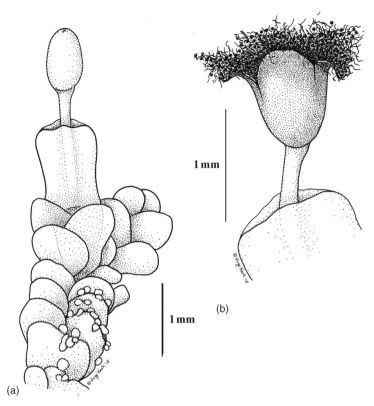

Fig. 3.14. The epiphytic *Radula complanata* reproduces by both vegetative, discoidal gemmae on the leaves and spores (a). Note the short seta that ensures that the capsule dehisces beyond the perianth. The capsule exhibits a typical four-valved dehiscence, liberating the spore mass and elaters (b).

differentiates into a single elater, which is a diploid elongated cell with spiral wall thickenings whose hygroscopic movements enhance spore dispersion. The sporocyte may, by contrast, undergo further divisions prior to meiosis. At maturity, spores may thus outnumber elaters by more than a factor of four.

The sporophyte is green and photosynthetic when young. The elongation of the seta pushes the capsule against the calyptra and causes it to tear. The calyptra thus remains completely attached to the maternal gametophyte and never forms a hood covering the sporangium as in mosses. Capsule dehiscence and spore release occur shortly after seta elongation ceases, often within a few hours of capsule emergence. In the vast majority of liverworts, the capsule dehisces along joint lines. In leafy and simple thalloid liverworts, normally two joint lines divide the capsule into four valves that separate from their apex down (Figs. 3.14b, 3.16). With drying, the cell walls between the two rows of sutures tear and the four valves bend backward, releasing the

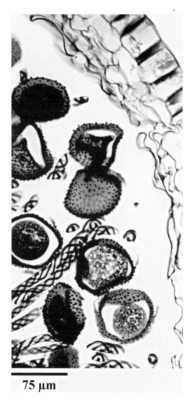

75 μm

Fig. 3.15. Spores and elaters in *Porella*, as seen in a transverse cross-section of the multistratose capsule wall (reproduced from Crandall-Stotler *et al.* 2009 with permission of Cambridge University Press).

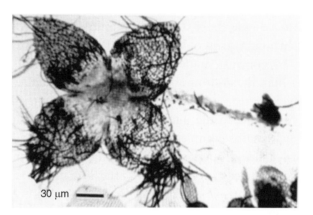

30 μm

Fig. 3.16. Four-valved dehiscence in *Jubula*. In the Jubulaceae and Lejeuneaceae, the elaters remain attached to the capsule valves after dehiscence (reproduced from Crandall-Stotler *et al.* 2009 with permission of Cambridge University Press).

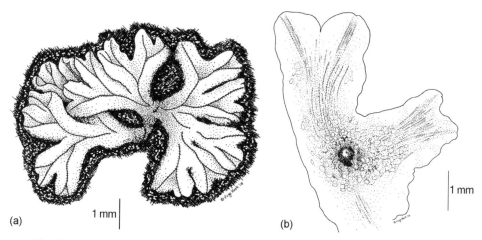

Fig. 3.17. *Riccia huebeneriana*. Habit of thallus (a) and close-up showing the sporophyte developing inside and emerging from the thallus following the disintegration of the superficial tissues (b).

spore mass and elaters. In complex thalloid liverworts, the dehiscence follows either a single (e.g. *Monoclea*) or several longitudinal lines, which can be either regular (e.g. *Lunularia*) or irregular. In some complex thalloid liverworts, such as *Asterella*, an apical lid, or operculum, is shed, as in mosses. Indehiscent capsules (cleistocarpy), although rare, characterize some complex (e.g. *Riccia*) or simple (e.g. *Fossombronia*) thalloid species (Fig. 3.17). The breaking-up of the spore mass and the dispersal of the spores is enhanced by hygroscopic movements of the elaters (Fig. 3.14b). In some cases, the elaters are attached at one or both ends to the capsule wall, preventing a premature dehiscence. The spores are typically unicellular and endospory, resulting from precocious divisions, occurs primarily among epiphytic leafy liverworts (i.e. Porellales). In the dioicous genus *Sphaerocarpos*, the two male and two female spores resulting from each meiosis remain attached in tetrads when dispersed, ensuring that both male and female gametophytes will co-exist.

3.1.3 Asexual reproduction

Reproduction in the absence of sex is, as in mosses and hornworts, common practice among liverworts. Although gametophytes may be generated from small fragments, asexual propagation in many taxa involves the development of specialized propagules. The diversity of innovations is astonishing (Fig. 3.18) and even congeneric species may differ in their diaspore type. Many species develop propagules along the leaf or thallus margins (Fig. 3.18a–d). In others, propagules arise from the leaf or thallus surface (Fig. 3.18e). In rare

Fig. 3.18. Diversity of vegetative propagules in liverworts. (a) Single-celled gemmae arising from leaf margins in *Nowellia curvifolia*. (b) Single-celled gemmae arising from leaf margins of *Lophozia capitata*. (c) Discoid, multicellular gemmae along leaf margins of *Radula australis*. (d) Polygonal gemmae from leaf margins of *Lophozia bicrenata*. (e) Multicellular gemmae arising from leaf surface in *Plagiochila*. (f) Gemmae cups in *Marchantia* (photos P. Davison). See plate section for colour version.

cases, they are formed internally within the cauline parenchyma or simply by the development of a new, likely resistant, cell wall within existing epidermal cells (Schuster 1984a). Dispersal of entire modules consisting of branches often with reduced leaves occurs in various Lejeuneaceae. *Marchantia* and *Lunularia*, both complex thalloid taxa, are well-known for their discoid gemmae aggregated in small cups on the surface of the thallus (Fig. 3.18f). The discs are stalked and dispersed by raindrops imploding in the cups. The gemmae are bipolar, with each pole generating a new gametophyte. Perennating structures known as tubers provide dormant stages essential for the gameto-phyte to aestivate. They are developed from the thallus and rarely from rhizoids, as in mosses or hornworts.

3.1.4 Symbiotic associations

About half of all liverwort species establish a close association with fungi and a majority of them seem obligately symbiotic, as they are only known in the infected state (Wang & Qiu 2006). The partnership leaves the host tissue typically intact, but whether it actually benefits the liverwort has rarely been demonstrated and is otherwise only assumed (Read *et al.* 2000). Most liver-worts harbour a phylogenetically diverse assemblage of glomeromycetes that often form the characteristic intracellular arbuscules (Fig. 1.12). Leafy liver-worts can also be colonized by ascomycetes or, more rarely, basidiomycetes (Duckett *et al.* 2006a). The association is not very specific, since a single liverwort species may be colonized by more than one fungal species. Further-more, phylogenetic ancestry does not constrain liverworts in their choice of fungal partner (Ligrone *et al.* 2007). In fact, experimental evidence suggests that liverworts may even adopt endophytes found on vascular plants and thus that co-evolution of fungal and liverwort partners is unlikely. The broad distribution of fungal associations across lineages resulting from the early diversification of liverworts (i.e. *Haplomitrium*, *Treubia* and complex thalloid species) suggests that the ability, and maybe even the need, to establish a partnership with fungi is a plesiomorphic trait among liverworts and, by extrapolation, for land plants (see Section 1.4.2). The fungus typically penetrates the thallus through the rhizoids (Russell & Bulman 2005). In *Haplomitrium*, which lacks rhizoids, the fungus directly penetrates the subterranean axes through the epidermal cells that secrete abundant mucilage, which is thought to promote fungal growth (Carafa *et al.* 2003). In essence, the fungal hyphae function as rhizoids to the *Haplomitrium* plants. Similarly, in *Treubia*, the sister taxon to *Haplomitrium*, the glomeralean fungus penetrates the plant not through the rhizoids, which remain fungus free, but through the ventral epidermis. Considering the phylogenetic antiquity of *Haplomitrium* and *Treubia*, these observations led to the hypothesis that fungal associations pre-date the origin of rhizoids in liverworts. Rhizoids were thus not required for the colonization by endophytic fungi, a hypothesis that would be congru-ent with the origin of endomycorrhizal fungi pre-dating the origin of land plants (Berbee & Taylor 2007).

Whereas liverworts commonly associate with fungi, symbiotic relationships with cyanobacteria are extremely rare and confined to members of the Blasiales. Both species of this lineage, *Blasia pusilla* and *Cavicularia densa*, harbour *Nostoc* colonies, which are chemically attracted (Adams & Duggan 2008), in ventral auricles. Within two weeks of innoculation, filaments protrude from the cavity wall into the colonies, probably to increase the surface area for exchange and in particular the mining of the nitrogen fixed in the heterocysts of the *Nostoc* endophyte (Rodgers & Stewart 1977). The liverwort–*Nostoc* association, which is not specific (Rikkinen & Virtanen 2008), must be established de novo following reproduction by spores, but not when establishing new individuals from gemmae, which often harbour the symbiont (Duckett & Renzaglia 1993).

3.2 Classification and macroevolution

Recent phylogenetic inferences from sequence data challenge morphology-based concepts of liverwort taxonomy and the recognition of three main lineages defined by the architecture of the gametophyte (Heinrichs *et al.* 2005, Forrest *et al.* 2006, He-Nygrén *et al.* 2006). Indeed, the simple thalloid liverworts are not monophyletic; the circumscription of the Marchantiales (the complex thalloid liverworts) has been revised; and it is clear that, although most leafy liverworts compose a well-supported clade, leaves evolved independently several times in the course of liverwort evolution. The consequence of this novel hypothesis is that many characters used tradi-tionally to define the three lineages are homoplastic and thus poor indicators of phylogenetic relationships. Furthermore, some lineages now lack morpho-logical characteristics. Exploring the biochemical, morphological and onto-genetic diversity of liverworts de novo in search of new synapomorphies that would support the shared ancestry of newly aligned taxa has therefore become the latest challenge for liverwort systematists.

The relationships among the major lineages of liverworts are summarized in Fig. 3.19. A detailed account of the phylogeny is presented by Crandall-Stotler *et al.* (2009). *Haplomitrium* and *Treubia* compose the sister group to the remainder of liverworts, within which the primary cladogenetic events separate the complex thalloid lineages from the simple thalloid and leafy

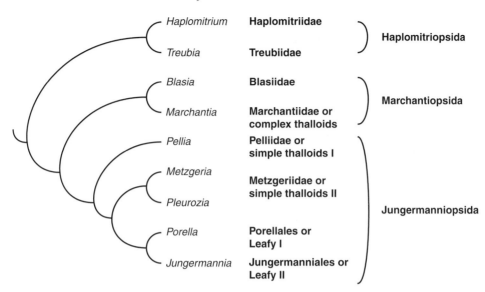

Fig. 3.19. Simplified liverwort phylogeny (redrawn from Forrest *et al.* 2006).

liverworts. Simple thalloid liverworts are now interpreted as a paraphyletic grade composed of the Pelliidae (also known as simple thalloids I) and the Metzgeriidae (simple thalloids II; Fig. 3.1d). The Pelliidae are thalloid and occasionally foliose (e.g. *Noteroclada*) and include taxa with a central strand containing dead water-conducting cells within the thallus mid-region (see Section 1.4.2). The Metzgeriidae lack any internal differentiation and comprise simple thalloid species forming a sister clade to the leafy *Pleurozia* (Fig. 3.20). This heterogeneous lineage composes the sister group to the remainder of the leafy liverworts, or Jungermanniidae, split in two lineages, Leafy I (or Porellales) and Leafy II (or Jungermanniales). The Porellales include *Porella*, *Radula* and the most diverse family of liverworts, the Lejeuneaceae. The Porellales are characterized by leaves with a small ventral lobule (Fig. 3.1i) and, except for *Ptilidium*, the endosporic germination of spores (Heinrichs *et al.* 2005). Incubous leaf insertions are shared by all taxa but are not diagnostic of the clade, since this character is also expressed in its sister group. The Jungermanniales comprise a diverse assemblage of taxa that lack any conspicuous shared morphological innovation. The leaves are entire or lobed (with the dorsal lobe smaller), the insertion is transverse and either incubous or succubous and branches arise ventrally or laterally.

A primary source of homoplasy is reverse evolution, whereby synapomorphic character-states are lost during the diversification of a clade. Character loss has occurred repeatedly within the complex thalloid taxa with varying degrees of severity (Boisselier-Dubayle *et al.* 2002). The genus *Monoclea*, for example, was

Fig. 3.20. *Pleurozia gigantea* looks like a leafy liverwort, but is nested within a clade of simple thalloid liverworts (photo L. Zhang). See plate section for colour version.

traditionally considered intermediate between the complex and simple thalloid liverworts: the gametophyte bears scattered oil cells and the antheridia are clustered on receptacles, as in complex thalloid liverworts, but the thallus lacks air chambers and pores and the seta is very long as in simple thalloid liverworts (Johnson 1904, Schuster 1984b). Analyses of sequence data as well as secondary metabolite chemistry (Gradstein *et al.* 1992) reveal that *Monoclea* is nested within a group of complex thalloid taxa. The lack of internal differentiation in the vegetative body is thus best interpreted as a reduction. *Blasia*, another taxon previously considered to have affinities to the simple thalloid taxa (Schuster 1984b), is now known to share a unique ancestor with the complex thalloid lineages. Whether this relationship indicates a derivation of the complex anatomy from a simple thallus, or an origin of *Blasia* through reduction, is not clear.

Parallel evolution of innovations is another source of homoplasy. The novel phylogeny inferred from DNA sequences suggests, for example, that the evolutionary history of liverworts is characterized by several transitions between thalloid and leafy organization. Whether leaves or a continuous thallus constitute the innovation that has arisen more than once is ambiguous, due to the uncertainty of the primary cladogenetic event at the base of the tree leading to thalloid (complex thalloid) and leafy taxa (*Haplomitrium* and *Treubia;* Fig. 3.1a, b). Recent phylogenetic reconstructions also reveal the shared ancestry of taxa with highly dissimilar morphologies. Most notable is the alliance of *Pleurozia* with Metzgeriidae II. *Pleurozia* has a leafy gametophyte with lobulate leaves (Fig. 3.20) that lacks any commonality with the simple ribbon-like thallus of *Metzgeria* and its relatives, except for the apical cell that is lenticular in shape (Fig. 3.2c), and likely inherited from a common ancestor. In all other leafy liverworts, the apical cell is tetrahedral (Fig. 3.2a, b). Although DNA data may be essential for accurately reconstructing the phylogeny of liverworts, continuous exploration of morphological and ontogenetic characters is imperative for interpreting the evolutionary trends and identifying the innovations defining the various lineages.

3.3 Biogeography and ecology

Liverworts are found literally everywhere. They flourish on every continent and landmass, including Antarctica, and exploit a remarkably diverse array of microhabitats. Except for the ocean and other salty aquatic habitats, liverworts can be found in the most extremely harsh environments. Some species, for example, initiate the process of colonization on bare volcanic rocks (see Section 2.2). Others thrive beneath dense moss carpets. They are not capable of photosynthesis and are, hence, dependent on green plants for carbon alimentation, a unique feature among bryophytes. In fact, *Cryptothallus* (Fig. 3.21) is specifically associated with fungi of the genus *Tulasnella*, which in turn forms mycorrhizae with *Betula* and *Pinus*. *Tulasnella* provides the pathway through which carbon flows from one of its autotrophic hosts to *Cryptothallus* (Bidartondo *et al.* 2003) (Fig. 3.22).

Although the large simple thalloid genus *Metzgeria* mostly includes epiphytic species, thalloid liverworts in general tend to occur on soil, whereas leafy liverworts occur on a wider range of substrata, including rock, bark and living leaves. Therefore, thalloid taxa dominate in open environments, whereas leafy liverworts are best represented in closed vegetations. On the Galapagos Islands, for example, thalloid liverworts are the only form found on bare soil in the pampas and are dominant near sea level in the arid lowland zone, whereas the trend is reversed in the dense, closed evergreen forest at middle elevations, where only leafy taxa occur (Gradstein & Weber 1982) (Fig. 3.23).

Leafy liverworts are particularly prevalent in moist and cool habitats. They are luxuriant in tropical rainforests, where they typically exhibit a higher species diversity than mosses owing to the contribution of some highly diverse families such as the Lejeuneaceae and Plagiochilaceae (Sillett *et al.* 1995, Holz *et al.* 2002, Holz & Gradstein 2005). Because leafy liverworts represent the bulk of species diversity among liverworts, they influence the global diversity patterns of the group as a whole. As a result, liverwort diversity is much higher in the areas of the world with a wet climate, such as the Andes, southern

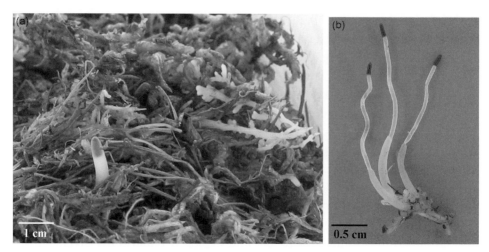

Fig. 3.21. The heterotrophic liverwort *Cryptothallus*. Whitish thallus with a sporophyte emerging from a thick *Sphagnum* carpet (a) and detail of a fertile thallus with three sporophytes (b) (photo N. Wickett). See plate section for colour version.

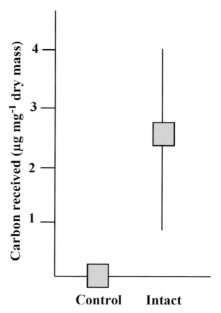

Fig. 3.22. Concentration of carbon received 21 days after a $^{14}CO_2$ pulse by individual thalli of the heterotrophic liverwort *Cryptothallus* grown with *Betula* (reproduced from Bidartondo *et al.* 2003 with permission of The Royal Society and the authors). In the control, hyphal connections between thalli and roots were severed immediately before $^{14}CO_2$ exposure.

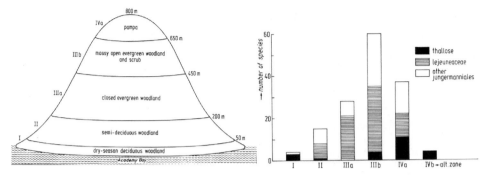

Fig. 3.23. Species richness in thalloid and leafy liverworts (divided into members of the family Lejeuneaceae and others) in the different vegetation zones along an altitudinal gradient in the Galapagos Islands (reproduced from Gradstein & Weber 1982 with permission of the Hattori Botanical Laboratory).

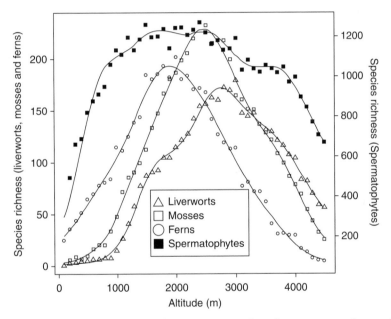

Fig. 3.24. Liverwort species richness compared to fern, moss and vascular plant species richness (i.e. total number of taxa reported every 100 m between 100 and 5000 m) along an altitudinal gradient in Nepal (reproduced from Grau *et al.* 2007 with permission of Blackwell).

South America and New Zealand, than in those regions with globally much drier climates, such as northern Africa and Arabia (Fig. 6.5). Because of their tendency to seek moisture and their ability to withstand a short growing season and low temperatures, the shift of the peak in species diversity commonly observed at mid-elevation along an altitudinal gradient occurs towards higher altitudes for leafy liverworts than for other plants (Fig. 3.24) (Grau *et al.* 2007).

Fig. 3.25. Thalloid communities from xerotropical *Sterculia africana* woodland in the Jabal Arays, Yemen. (a) Location of the communities within the
landscape. (bi) *Riccia argentolimbata*. (bii) *Riccia atromarginata*. (biii) *Exor-
motheca pustulosa*. (c) *Mannia androgyna*. (d) Detail of *Riccia atromarginata*
(reproduced from Kürschner 2003 with permission of Gebrüder Borntraeger).
See plate section for colour version.

Thalloid liverworts, by contrast, globally tend to occur in more xeric environments. They predominate in Mediterranean and xerotropical environments,
where they may form dense, gregarious patches on bare, very thin soil layers
(Fig. 3.25). The 'xerothalloid' life syndrome refers to the strategy displayed by

characteristic genera of those habitats. Xerothalloids include annual plants, such as *Riccia* (Fig. 3.25), whose gametophyte dies during the dry season. Species thus persist as spores, which are typically fairly large (50–100 μm in diameter) and produced within cleistocarpous capsules enclosed in the thallus. Upon the decay of the gametophyte and the capsule wall (Fig. 3.17), spores are released in the immediate vicinity of the former parental gametophyte (Kürschner 2003). Xerothalloids are also represented by perennial species whose thallus is adapted to drought. In the genus *Targionia*, for example, the margins of the thallus roll up on drying, thereby protecting the green dorsal side of the thallus and exposing a ventral side covered by blackish scales (Fig. 8.12).

4

Mosses

Mosses (Division Bryophyta) are generally seen as small plants confined to humid habitats, avoiding exposure to direct sunlight. Yet, an alert naturalist will quickly notice their presence in virtually every ecosystem. In parts of the world where short growing seasons limit plant growth, mosses may dominate the vegetation. Similarly, in temperate and tropical rain forests, mosses compose luxuriant epiphytic communities that play important ecological functions, especially in terms of water and nutrient flow. Approximately 12 000 species of mosses are currently recognized, reflecting a broad morphological diversity. In fact, a unifying macroscopic definition is impossible due to both fundamental architectural differences among major lineages and extensive reduction, and hence character loss, across the phylogenetic tree of mosses. A sporophyte composed of a robust stalk elevating the capsule diagnoses most mosses, but is lacking in peat mosses and various ephemeral taxa inhabiting seasonally dry ecosystems. In many cases, closer examination further reveals one or two rings of teeth lining the capsule mouth. The vegetative (gametophyte) body always consists of a terete axis with sessile leaves, rather than a flattened thallus as in hornworts and some liverworts. Although associations with fungi are common, evidence for a symbiotic nature of the relationship benefiting the moss is lacking.

4.1 Structure and development

4.1.1 Gametophyte

The architecture of the vegetative body of mosses is fundamentally modular: it is built from blocks of cells assembled into axes and multiple axes are joined to form a stem or branch system. The series of divisions that gives rise to all cells composing an axis typically begins with the single apical cell dividing

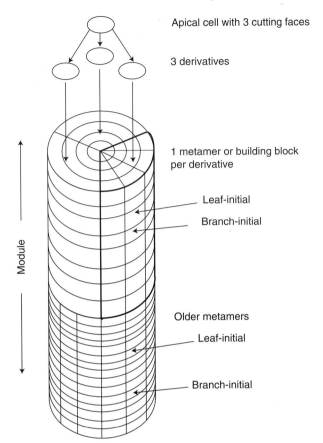

Apical cell with 3 cutting faces

3 derivatives

1 metamer or building block per derivative

Leaf-initial

Branch-initial

Older metamers

Leaf-initial

Branch-initial

Module

Fig. 4.1. Modular architecture of the moss shoot. The apical cell undergoes divisions parallel to its three cutting faces. Each derivative undergoes a series of divisions and the resulting cells form a building block or metamer. Each of these holds a leaf- and a branch-initial that may develop into a single leaf or branch, respectively. All metamers along a single axis compose a module. The module can be a stem or a branch and is typically foliated.

along each of its three cutting faces to yield derivatives (Fig. 4.1). These daughter cells then undergo a suite of divisions. The resulting blocks of cells, also called metamers, are assembled into a module, which is the axis. Apical growth in mosses, like in other bryophytes, thus generally results from the activity of single apical cell rather than a multicellular meristem as in vascular plants. The module is always terete and typically bears leaves. The gametophyte is never thalloid at maturity and only in *Sphagnum* and *Andreaea* are dorsi-ventral thalli or protonemal appendages developed.

Growth of the moss gametophyte results from the accumulation of successive modules, unless the plant is composed of a single module and does not branch. Each building block extends from the surface to the centre of the axis, and includes two superficial cells destined to form a leaf or a branch, respectively. The branch-initial may undergo a series of divisions forming a primordium or bud prior to entering dormancy. Branching thus results from the development of a lateral branch-initial and never from the splitting of the apical cell. Branching in mosses is, therefore, never dichotomous. Some mosses remain unbranched, whereas others develop branches only at the apex of the stem or more or less regularly along the main axis. The architecture of branches is identical to that of the stem: they grow from an apical cell and building blocks formed by successive divisions of the derivatives compose the new axis. It is perhaps this type of modular growth, merely consisting of adding new blocks to existing ones, that renders mosses and, in fact, other bryophytes, so prone to morphological changes when habitat conditions change (Box 4.1).

Branches typically bear leaves, which arise from leaf-initials. Below each leaf-initial is a branch-initial, so that branches can also branch! The distribution of branches is determined by the dormancy of the initials. Some initials remain dormant throughout the life of the plant, others immediately develop into a branch. In mosses, wherein the formation of sex organs at the apex of the stem consumes the apical cell, vegetative growth may only resume through the activation of a branch-initial to form a new branch or innovation. This branching pattern is referred to as sympodial branching, and characterizes many acrocarpous mosses (i.e. those with terminal female sex organs). By contrast, pleurocarpous mosses branch monopodially, maintaining apical growth of their main module and developing branches independently of sexual maturation. In fact, the development of sex organs is transferred to highly differentiated branches that remain short and bear leaves primarily surrounding the gametangia. In cladocarpous mosses, the branches are all well developed and most complete their development with the formation of female sex organs. Such organization is likely to have arisen multiple times from an acrocarpous condition. Variation in the mode of branching of different modules (sympodial or monopodial), the polarity of branch development (acropetal or basipetal), the direction of growth (orthotropic or plagiotropic) and the density of branching (determined by the dormancy of successive initials), yields a broad diversity of body plans or life forms in mosses (La Farge-England 1996; Figs. 4.2–4.3). Peat mosses (members of the genus *Sphagnum*) are easily diagnosed by their fasciculate branches (Fig. 4.2b): only every third or fourth branch-initial develops into a branch and each of these rapidly produces new innovations. Branches typically perform the same function as stems, carrying the photosynthetic leaves. In cladocarpous and pleurocarpous mosses, the branches also produce the terminal sex organs. Even in monoicous acrocarpous mosses, branches

Box 4.1
Plants vary

Plant species, like any other organism, vary. In bryophytes, different populations or individuals often exhibit a wide range of morphological variation. Alterations of morphological characters are particularly pronounced along a moisture gradient. The moss *Climacium americanum*, for example, varies in growth form from erect and strongly dendroid to elongate and trailing with an irregular branching pattern (Box 4.1 Fig. 1) (Crum & Anderson 1981). Such a variation is correlated with habitat conditions: plants are generally dendroid in mesic forests

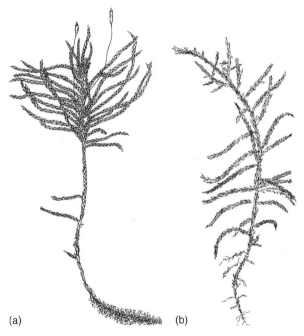

(a)   (b)

Box 4.1 Fig. 1. *Climacium americanum*, habit of a plant from mesic forest habitats (a) and from a swamp (b) (reproduced from Crum & Anderson 1981 with permission of Columbia University Press).

and are trailing in creeks and swamps. The main question is how to determine whether such a variation has a genetic basis or is due to phenotypic plasticity, i.e. is simply reversibly induced by local environmental conditions. While the *Climacium* case was, at first sight, suggestive of plasticity, molecular markers revealed that the morphological differences were actually genetically controlled (Shaw *et al.* 1987). In fact, the respective influence of the genome and of the environment in shaping morphological variation are difficult to disentangle and the causes of morphological variation in bryophytes remain largely unexplored.

Box 4.1 (cont.)

Cultivation under controlled conditions consistently reveals, however, that bryophyte species are extremely plastic. In general, alterations pertain primarily to the density of branching and foliation, the shape of the leaves, the thickness and length of the costa and the shape and size of the cells (Box 4.1 Fig. 2). Morphological transformations may also concern plant anatomy. For example, a moist sample of *Sphagnum fimbriatum* left in a cold room under low light intensity develops new shoots with leaves lacking the characteristic alternation of hyaline and chlorophyllose cells. In many mosses, the differentiation of water-conducting cells in the stem may also be weakened or even become absent in moist conditions (Hébant 1977).

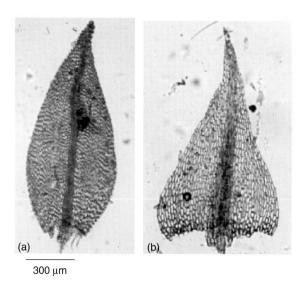

(a)   (b)

300 µm

Box 4.1 Fig. 2. Plasticity of stem leaf shape in *Hygroamblystegium*: oval shape with obtuse apex, characteristic of *H. fluviatile*, from the wild (a) and triangular shape with an acuminate apex after three months of cultivation in a phytotron (b), resulting in a leaf morphology that is actually typical of *H. tenax* (reproduced from Vanderpoorten 2004 with permission of the Missouri Botanical Garden).

Bryophytes may also vary in the course of their development, from the protonema to the fully developed gametophytic stage. While this is rarely an issue since the protonema of most mosses is not persistent, an extreme example was most recently provided by the discovery of an aquatic algae-like green mat-forming plant composed of densely branched filaments, each one cell wide, in several rivers in Japan (Box 4.1 Fig. 3) (Higuchi *et al.* 2003). Analysis of a partial chloroplast gene sequence and cultivation on an appropriate medium

Box 4.1 (cont.)

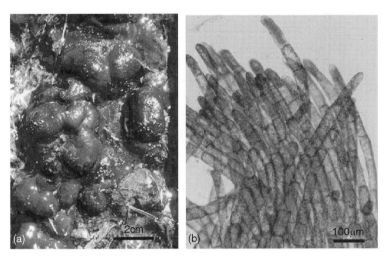

Box 4.1 Fig. 3. General aspect (a) and close-up of the filaments (b) of the permanent moss protonema found at the bottom of rivers in Japan (reproduced from Higuchi *et al.* 2003).

revealed that the plant was actually a moss, closely related and potentially a member of the genus *Dicranella* (Dicranaceae).

Bryophytes are thus ontogenetically and morphologically labile. The resulting variation in characters is the nemesis of the taxonomist. For the evolutionary biologist, it raises the question of the significance of such fundamental plasticity.

arising below the apex may rapidly cease vegetative growth to form a cluster of gametangia of the sex complementing the one on the main stem. In several species, branches serve as propagules for asexual reproduction: they bear reduced leaves and are deciduous.

The anatomy of the stem or branch is typically simple, consisting of a mostly unistratose epidermis that always lacks stomata, a cortex of parenchymatous cells and, in some cases, a central strand of water-conducting cells (Fig. 4.4a). The epidermal cells are often narrower than the cortical cells and their wall is distinctly thickened, at least relative to the size of the cell lumen. The walls are typically pigmented, giving the stem a reddish coloration. In a few mosses, primarily hygrophytes, the epidermal cells are thin-walled and somewhat inflated, forming a so-called hyalodermis, which may function in external water conduction. Many pleurocarpous mosses produce paraphyllia (Fig. 4.5),

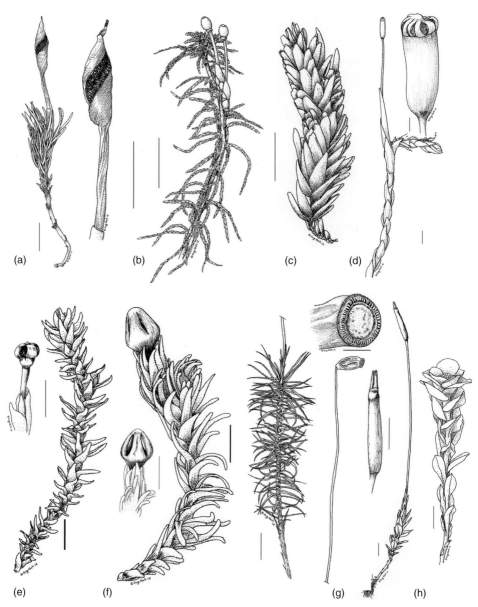

(a)   (b)   (c)   (d)

(e)   (f)   (g)   (h)

Fig. 4.2. Moss diversity. (a) *Takakia ceratophylla*: the leaves are deeply lobed and the capsule dehisces along a spiral line. (b) *Sphagnum fimbriatum*: the branches are arranged in fascicles and the capsule is elevated on a pseudo-podium, rather than a seta. (c) *Ambuchanania leucobryoides*: although closely related to *Sphagnum*, the branches are not grouped in fascicles and the architecture of the leaves is also distinct. (d) *Acroschisma wilsonii* belongs to the Andreaceae, but the capsule dehisces only near the apex, with lines often converging. (e) *Andreaea rupestris* dehisces along four lines but the valves remain attached at their apex; the capsule, which is carried by a

slightly branched photosynthetic epidermal outgrowths that may also contribute to the movement of water over the stem surface. The complexity of these paraphyllia varies in terms of degree of branching, size and surface ornamentation, offering characters suitable for distinguishing species. Unbranched hyaline filaments are found only in leaf axils, hence their name, axillary hairs (Fig. 4.5). They arise from the leaf-initial and secrete a mucilage that prevents the dehydration of the developing organ. The hairs are deciduous and are best observed in the axil of juvenile leaves. They vary greatly in size, shape and number among but not within species and hence may be informative for distinguishing taxa. The epidermis also develops small photosynthetic appendages, called pseudoparaphyllia (Fig. 4.5), that are confined to the periphery of the branch primordium that they protect.

The body of the stem is composed essentially of undifferentiated, broad, elongated and non-photosynthetic parenchyma cells. These cells may store lipids or starch, or serve a structural function, in which case their walls are thickened. In many taxa, the axial cells are differentiated to form a central conducting strand (Fig. 4.4a). These cells, called hydroids, are much narrower, highly elongated with tapered ends (except in *Takakia*), and lack cytoplasmic content at maturity (see Section 1.4.2). Conducting strands are lacking in *Sphagnum*, *Andreaea* and *Andreaeobryum*, and are best developed in the Polytrichopsida. Photosynthates are distributed either through parenchyma cells or specialized cells called leptoids that are also best developed in the Polytrichopsida (see Section 1.4.2). In addition to their ultrastructural specialization, these cells differ markedly from their neighbours in the high frequency of plasmodesmata in the oblique cross walls (Ligrone *et al.* 2000). Some mosses develop multiple water- and food-conducting strands in the stem but, except in the Polytrichopsida, these vanish before reaching the base of the midrib of the leaf.

The gametophyte is anchored to the substrate by multicellular and generally uniseriate rhizoids. These arise from the epidermal cells of the stem or branches, or even from the costa. Rhizoids are absent in *Takakia* and restricted to the

Caption for Fig. 4.2. (cont.)
pseudopodium, contracts upon drying, exposing the spore mass. (f) *Andreaeobryum macrosporum* resembles species of *Andreaea* but has a seta. (g) *Polytrichum commune*: the capsule bears a nematodontous peristome composed of 64 teeth. (h) *Tetraphis pellucida*: the peristome consists of four massive teeth, composed of entire cells; plants may also form terminal gemmae cups, as shown on the right. Scale (a, c, d, e) (incl. capsule close-up), (f) (incl. capsule close-up), (g) (capsule close-up only) and (h) = 1 mm; (b) = 1 cm; (g) = 5 mm.

Fig. 4.3. Moss diversity (continued). (a) *Buxbaumia aphylla*: the vegetative phase is composed of a persistent protonema, the gametophyte is reduced but the sporophyte is massive. (b) *Diphyscium foliosum*: similar to *Buxbaumia*, although not most closely related; the capsule is sessile, barely emerging from among the leaves. (c) This undescribed species is characterized by a large indehiscent capsule and numerous stomata. (d) *Gymnostomiella vernicosa*: a tiny moss reaching 2 mm. (e) *Tayloria gunnii*: a typical dung moss, with the sterile base of the capsule much enlarged (see Box 4.2). (f) *Orthotrichum rupestre*: this acrocarpous perennial moss shows sporophytes from previous years. (g) *Climacium dendroides*: the branches are clustered at the tip of the stem, and the plants are reminiscent of trees, hence the name 'dendroides'. (h) *Plagiothecium undulatum*: this plagiotropic moss often develops leaves that lie in a single plane. (i) *Thuidium delicatulum*: a typical pleurocarp, with a stem producing abundant branches at regular intervals. (c and d) previously published by BG in the Flora of Australia vol. 51 mosses 1. Scale (a) = 5 mm; (b, c, e) = 1 mm; (f) = 2 mm; (g) = 2 cm; (d, h, i) = 0.5 mm.

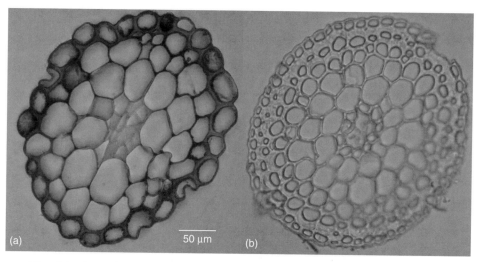

Fig. 4.4. Stem (a) and seta (b) anatomy in *Funaria hygrometrica*.

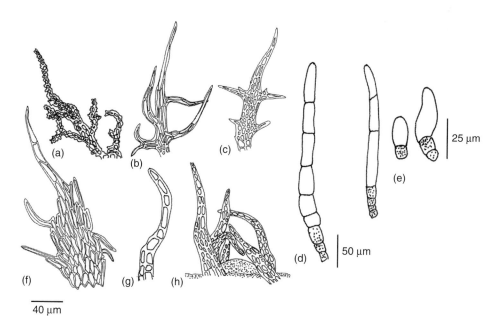

Fig. 4.5. Stem appendages. Paraphyllia (a–c). Axillary hairs (d, e). Pseudo-paraphyllia (f–h). (a) *Thuidium delicatulum*. (b) *Hylocomium splendens*. (c) *Thelia hirtella*. (d) *Catharomnion ciliatum*. (e) *Cyathophorum adianthum*. (f) Foliose pseudoparaphyllium in *Hypnum imponens*. (g) Filiform pseudo-paraphyllium in *Pterigynandrum filiforme*. (h) Pseudoparaphyllia surrounding branch primordium in *Neckera complanata*. (d, e) redrawn from Kruijer (2002); (a, b, c, f) redrawn from Crum & Anderson (1981); (g, h) redrawn from Ireland (1971).

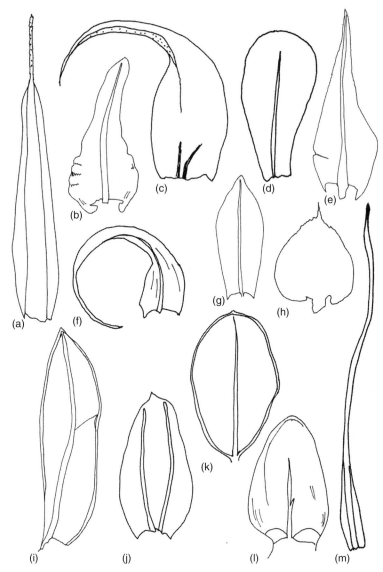

Fig. 4.6. Diversity in shape of vegetative leaves. (a) Oblong-lanceolate leaf with a broad costa occupying more than half the width in *Campylopus pilifer*. (b) Oblong-ligulate leaf with an auriculate base in *Anomodon rugelii*. (c) Circinate, long tapered leaf from an ovate-lanceolate base with a short and double costa in *Hypnum hamulosum*. (d) Oblong-obovate leaf with broadly rounded apex and a slender costa ending in the upper third of leaf in *Gymnostomiella orcutii*. (e) Ovate-lanceolate leaf with a costa ending in an acute apex in *Meesia triquetra*. (f) Strongly falcate, long subulate leaf from a lanceolate base in *Sanionia uncinata*. (g) Oblong-ovate leaf in *Pyramidula tetragona*. (h) Broadly ovate leaf ending in a long apiculus and lacking a costa in *Myurella sibirica*. (i) Leaf with vaginant lamina (lower right),

protonemal stage in *Sphagnum*. Some pleurocarpous taxa also lack rhizoids on the mature leafy plant, although they may be developed in culture. Their colour varies from whitish in some species to brown or purple in others. Their surface may be either smooth or roughened by small papillae. Rhizoids are typically thigmotropic, winding around tiny objects. In some Polytrichaceae, narrow rhizoids even wind around larger ones, forming rope-like bundles. Rhizoids are not confined to the base of the stem in orthotropic species. In some Mniaceae, highly branched rhizoids may originate around the branch-initial, and hence serve a protective function. In other species, rhizoids cover much of the stem. The numerous capillary spaces in this spongy coating may be essential for external conduction or for holding excess water needed to delay dehydration of the plant and thereby prolong photosynthetic activity.

The leaves of mosses are always sessile: they lack a petiole and the lamina is thus inserted directly on the stem at its base (Fig. 4.6). Except for the deeply incised leaves of *Takakia* (Fig. 4.2a), moss leaves are never lobed. In *Fissidens* and *Bryoxiphium*, the photosynthetic area is increased by outgrowths of the midrib. The arrangement of leaves on the stem typically follows three parallel, spiralling lines. Some mosses, most notably *Fissidens*, display a distichous (i.e. two-ranked) arrangement of leaves. Leaves are generally distributed evenly along the axis, but, in some species, apical metamers fail to elongate, and the leaves are crowded into rosettes. The insertion of the leaf is typically transverse. Oblique insertions often result in leaves fanning out on both sides of the stem, seemingly laying in a single plane. Such complanate foliation characterizes several pleurocarpous taxa with prostrate stems, growing in low light environments (e.g. species of *Plagiothecium*; Fig. 4.3h).

All leaves on a single plant may be identical. Most mosses display, however, heteroblastic leaf development, with the leaves at the base of the module often lacking the modifications characteristic of the mature leaves found higher up the axis. Plants branching monopodially often bear stem and branch leaves that are dissimilar in size and also shape. Dimorphism also

Caption for Fig. 4.6. (cont.)
bordered all around in *Fissidens bryoides*. (j) Oblong (somewhat lingulate) leaf with a double costa ending near an apiculate apex in *Schizomitrium pallidum*. (k) Elliptic leaf with a narrow base, a rounded apex and a conspicuous border of elongate cells in *Rhizomnium punctatum*. (l) Ovate-elliptic leaf from a narrow base, with conspicuously differentiated alar cells in the corners, and ending in an obtuse apex in *Calliergon obtusifolium*. (m) Lanceolate leaf with a strong costa extending beyond the lamina in *Ditrichum flexicaule* (redrawn from Crum & Anderson 1981).

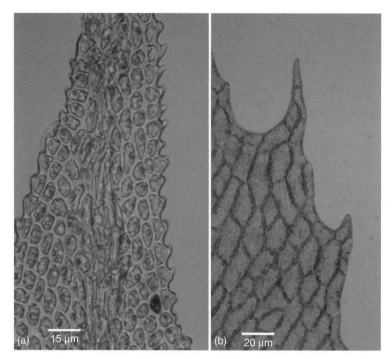

Fig. 4.7. Leaf margin. (a) Dentate due to projecting papillae. (b) Toothed, with multiple cells composing each tooth.

marks the leaves around the stem of a few species, with either the ventral or dorsal leaves being smaller than the lateral ones. The general shape of the leaf varies greatly among mosses, from nearly rounded, or tongue-shaped, to narrow and nearly linear (Fig. 4.6). The leaf is often entire, but lateral projections from either the apex or the sides of marginal cells confer dentations or serrations to the outline in many species (Fig. 4.7).

A single layer of cells composes the blade of a typical moss leaf, except for the median axis, which is often thickened to a multistratose midrib (or costa), a feature never expressed in leafy liverworts. This midrib is lacking in the basal lineages (i.e. those pre-dating the origin of *Oedipodium*; Fig. 4.22) and has also been lost in various derived lineages. Many pleurocarpous mosses have leaves with a double costa. The function of the costa may be two-fold: to provide structural support for the blade and to facilitate water and photosynthate transport to and from the leaf. Transverse sections of the costa reveal the presence of rather narrow and highly thickened cells, the stereids, which form distinct bundles or bands on one or both sides of the broader conducting cells, called guide cells. The midrib of many acrocarpous species is very prominent and in some it even occupies much of the width of the leaf.

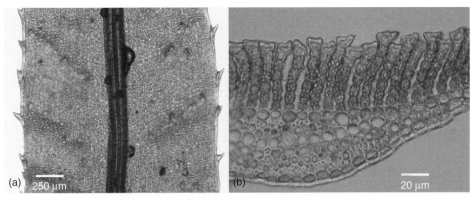

Fig. 4.8. Longitudinal photosynthetic lamellae on ventral surface of the costa of leaves in Polytrichaceae. (a) In *Atrichum*, the costa is narrow and mounted with few lamellae (here three; see dark vertical bands). In *Polytrichum*, the costa is much broader and the lamellae are numerous (b).

Lamellae of green cells mounted longitudinally on the ventral surface of the costa characterize various taxa, primarily within the Polytrichaceae and the Pottiaceae (Fig. 4.8). Wax covering the margins of the lamellae prevents water from filling the space between them, creating narrow air chambers. These chambers create reservoirs of CO_2, ensuring that the plant can maintain photosynthesis as water is moved into the leaves through the conducting cells.

Each leaf develops from one superficial leaf-initial, which quickly differentiates a single apical cell. This cell, however, soon ceases to divide and only its derivatives contribute new cells to the blade. Cell shape varies among lineages, between juvenile and mature leaves and even across the lamina. The laminal cells of acrocarpous mosses are typically five- to six-sided, isodiametric to short-rectangular (Fig. 4.9a, b). In contrast, those of many pleurocarpous taxa are greatly elongated and nearly linear (Fig. 4.9c). The cells of juvenile leaves are often uniformly thin-walled, lax, and without superficial ornamentation. The basal and marginal cells may be conspicuously differentiated from the median and apical cells (Fig. 4.9b). Basal cells are often elongate rectangular and, in some cases, enlarged and even highly inflated at the angle of the insertion (Fig. 4.9c). Sleeves of cells, or decurrencies, extending downward on the stem, are formed in many species. They may reinforce the attachment of the leaf or play an important role in external water conduction. Modification of the marginal cells into a distinct border occurs in various lineages of mosses (Fig. 4.9b). The border, often composed of two or more layers of cells, increases structural support for the leaf or functions in water transport to the upper cells. The remainder of the lamina is composed of chlorophyllose cells whose shape is often diagnostic of the

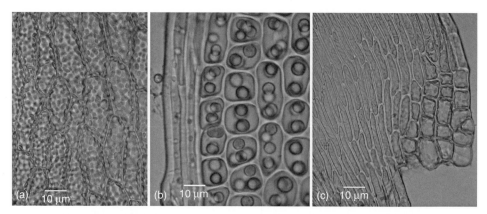

Fig. 4.9. Shape of laminal cells. (a) Six-sided. (b) Quadrate to isodiametric; note the linear marginal cells on left. (c) Elongate to linear; note the quadrate, slightly enlarged cells at the angle of the leaf base.

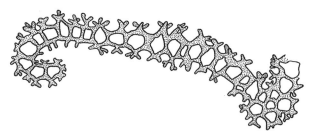

Fig. 4.10. Xerophytic taxa often develop conspicuous papillae on one or both surface of their laminal cells, as seen in a leaf cross-section of *Orthotrichum freyanum* (redrawn from Goffinet *et al.* 2007).

species. Cells may have highly thickened walls with, in some species, cytoplasmic connections. In many cases, the most striking feature of the leaves is the surface ornamentation. Laminal cells, particularly of species growing in drier habitats, typically bear projections on one or both surfaces (Fig. 4.10). These papillae, as they are called, confer a dull shine to the leaf. Their function may be to increase the absorption surface, thereby accelerating hydration when water is available, and also to create capillary spaces to hold excess water, delaying dehydration and prolonging photosynthesis. Through their apices, covered in water repellant wax, papillae also provide entry points for CO_2 and thereby enable photosynthesis to continue in times of water abundance. The papillae vary in shape from tiny hollow or solid warts to branched projections.

The most conspicuous dimorphism of cells within a single leaf is perhaps expressed in *Sphagnum* and *Leucobryum*, genera in which the leaves are

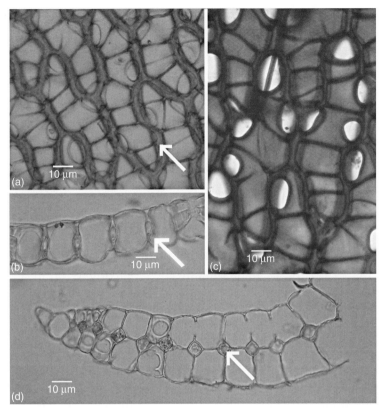

Fig. 4.11. Hyalocysts in *Sphagnum* (a–c) and *Leucobryum* (d). (a, b) Leaves of *Sphagnum* are one cell thick, with photosynthetic cells (arrow) alternating with hyaline cells or hyalocysts. (c) Hyalocysts of *Sphagnum* typically bear conspicuous pores, which are best revealed when the leaf is dyed with gentian violet. (d) In *Leucobryum*, seen here in leaf cross-section, hyalocysts form continuous layers enclosing the photosynthetic cells (arrow).

composed of a few, small, photosynthetic cells and large, hyaline, dead cells, the hyalocysts (Fig. 4.11). Such dimorphism is best explained by the need for the plants to acquire CO_2 while being hydrated. Water slows the diffusion rate of CO_2 and hence, while essential for photosynthesis, prevents the efficient assimilation of carbon. In *Sphagnum*, hyaline and green cells alternate in a single layer (Fig. 4.11a, b). The former hold water that diffuses into the photosynthetic cells, whereas the latter acquire carbon through their free surfaces. In *Leucobryum*, the photosynthetic cells form a layer sandwiched between two or more layers of dead cells (Fig. 4.11d). Hyalocysts in young leaves at the apex of the stem may serve as air tanks, whereas in older leaves they hold water.

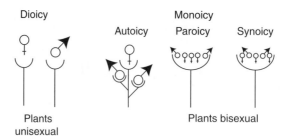

Fig. 4.12. Distribution of sex organs. Dioicy: male and female sex organs occur on different individuals and hence these plants are unisexual. Monoicy: both sexes occur on a single plant (although not always at the same time). Their distribution on the plant varies between autoicous, paroicous and synoicous species (see text for details).

Individual mosses may produce either both sexes (monoicy) or only one (dioicy; Fig. 4.12). Sexes in hermaphroditic individuals occur in distinct clusters (autoicy) or are grouped together, in which case the male sex organs may be surrounding (paroicy) or be intermixed (synoicy) among the female gametangia. Sexual dimorphism is uncommon, but extreme in phyllodioicous mosses, in which the male plants are dwarf, consisting of only a few leaves around the sex organs, and grow exclusively as epiphytes on the more robust female plants. Phyllodioicy has evolved multiple times and characterizes many tropical Orthotrichaceae, Leucobryaceae and Dicranaceae. Antheridia and archegonia originate at the apex of the stem or branch, forming conspicuous terminal clusters. In some mosses, isolated archegonia are formed in the axils of leaves. Groups of gametangia are typically surrounded by differentiated leaves that protect the delicate sex organs and also the young embryo. Clusters of archegonia and their associated leaves are called perichaetia, whereas similar male structures are referred to as perigonia. In some mosses, the leaves around the antheridia form a splash cup: sperm cells that are released in the water retained within the cup are dispersed by imploding raindrops. Sterile filaments, the paraphyses, are often mixed among the gametangia (although never in *Sphagnum*) and in some species they occur only with one sex or the other. They may be filled with mucilage, which retains water, and thereby protect the gametangia from dehydration. They may also secrete a chemoattractant for microarthropods recruited to disperse the otherwise dispersal-limited motile sperm (Harvey-Gibson & Miller-Brown 1927, Cronberg *et al.* 2006a).

As in other land plants, the gametangia consist of a multicellular jacket enclosing the egg or sperm cells. The archegonium is stalked and consists of a slightly inflated venter harbouring a single egg and mounted by a narrow

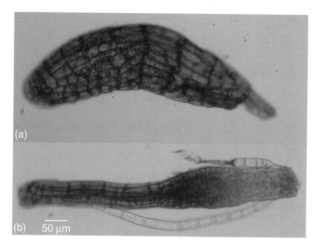

Fig. 4.13. Gametangia of mosses. (a) Antheridium. (b) Archegonium.

neck whose central canal cells disintegrate at maturity (Fig. 4.13b). The canal opens to the outside when the apical cells swell, as their mucilage absorbs water, and burst. The rather cylindrical antheridium is also stalked but contains numerous sperm cells (Fig. 4.13a). The moss antheridium differs from its homologues in liverworts and hornworts by the differentiation of opercular cells, which form a lid that is shed as the pressure in the lower part of the antheridium increases.

4.1.2 Sporophyte

The fundamental architecture of the moss sporophyte is simple, similar to that of liverworts: an unbranched seta anchored into the maternal gameto-phyte by the foot and carrying a single terminal sporangium. Although an apical cell is differentiated early during embryogenesis, its activity is ephem-eral. Soon, a new meristem, unique among bryophytes, is initiated below the apex. The segments above it will develop into the capsule and those below into the seta and the foot. Apical growth is thus confined to the earliest ontogenetic stages and much of the growth of the sporophyte results from the activity of the intercalary meristem. The foot is either tapered or bulbous and the placental region lining gametophytic tissues is typically composed of transfer cells. Whether the haustorium-like tissue is essential for the efficient uptake of water and nutrients, or merely for tightening the attachment of the sporophyte, is not clear.

The seta serves to raise the capsule above the perichaetial leaves protecting the developing sporophyte. The embryo is enclosed completely at first in

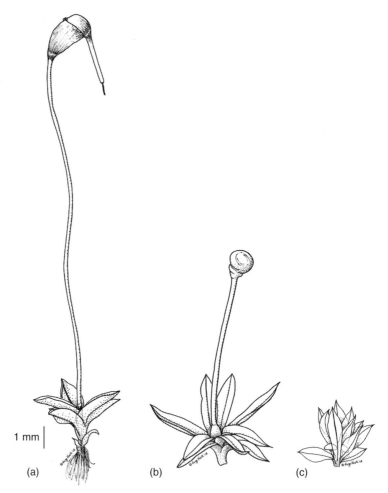

1 mm

(a)   (b)   (c)

Fig. 4.14. Sporophyte reduction in the Funariaceae. (a) Long exserted, curved capsule in *Funaria hygrometrica*. (b) Exserted, erect capsule in *Physcomitrium pyriforme*. (c) Immersed capsule (here the casule is already dehisced and remains as small cupules among the perichaetial leaves) in *Aphanorrhegma serratum*.

an epigonium that is derived from the archegonium. Occasionally, growth is halted prematurely and the capsule is retained within the perichaetium (e.g. *Aphanorrhegma*, Fig. 4.14). An intercalary meristem and thus an actual seta is lacking in only two genera, *Sphagnum* and *Andreaea*. In these lineages, elevation of the capsule is accomplished by growth of gametophytic tissue into a stalk called a pseudopodium. Lateral meristems and thus appendages are always absent from the seta and the epidermis that is composed of smooth or more rarely papillose cells universally lacks stomata. The stalk is solid and composed of parenchyma cells (Fig. 4.4b). Long, thin-walled water-conducting

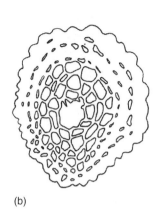

(a) 250 µm   (b)   (c) 300 µm

Fig. 4.15. Modification of the seta. In *Rhachitheciopsis*, the seta is twisted when dry (a) but erect when moist. The movement of the seta is accounted for by uneven wall thickenings of the cortical cells of the seta (b). In *Microbryum curvicolle*, an ephemeral ground-dwelling moss, the seta is curved downward (c), favouring dispersal of spores within the vicinity of the parental sporophyte (photo C. Rieser).

cells form an axial strand in some taxa, even those lacking conducting cells in the gametophyte. The walls of the cortical cells may be unevenly thickened, which leads to the twisting of the seta, even conspicuous bending in some taxa, upon drying, facilitating spore dispersal (Fig. 4.15a, b). In many, primarily ground-dwelling species, the seta is curved at the apex and the capsule hangs downward (Fig. 4.15c).

Early in the ontogeny of the sporangium, two tissues are differentiated. The four inner central cells form the endothecium. Its development lags behind that of the amphithecium, which develops from the outer rings and undergoes repeated divisions yielding new layers of multiple cells. The amphithecium forms the capsule wall and the underlying parenchyma as well as, in the region above the annulus, the teeth around the capsule mouth (i.e. the peristome). The endothecium forms the columella and the spore sac, although in *Sphagnum* this role is taken by the amphithecium. The columella is a sterile axis that extends through the sporogenous region and connects to the operculum except in basal taxa such as *Sphagnum* and *Andreaea*, in which it is dome-shaped, or in *Archidium*, which lacks a columella altogether.

The capsule typically bears stomata, dehisces subapically through the loss of a lid and bears one or two rings of teeth lining the mouth. Stomatal guard cells are rather elongate and kidney-shaped. They define a pore that is either round or slightly elliptical, resulting in a stoma that resembles that of other

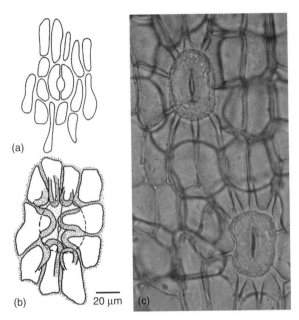

(a)

(b)   20 µm   (c)

Fig. 4.16. Stomata. (a) Phaneroporous: the guard cells are fully exposed on the surface of the capsule. (b) Cryptoporous: the guard cells are overarched by subsidiary cells. (c) Incomplete division of the guard cells results in the pore being defined by a single ring-shaped guard cell in the Funariaceae.

embryophytes (Fig. 4.16a). Occasionally, the mother guard cell fails to complete its division and the stoma is then defined by a single ring-like cell. Single guard cells are found in all members of the Funariaceae (Fig. 4.16c). Stomata are often restricted to the neck of the capsule. In some taxa, the guard cells are overarched by adjacent or subsidiary cells (Fig. 4.16b), which probably reduce airflow and thus evapotranspiration. Like stomatal crypts in vascular plants, such cryptoporous stomata are homoplasious and are found primarily in taxa growing in xeric conditions, such as many epiphytic or saxicolous members of *Orthotrichum*. Superficial or phaneroporous stomata are the general condition in mosses. Shortly after their differentiation, the stomata regulate gas exchange (Garner & Paollilo 1973). They remain open thereafter and their function remains ambiguous. In *Sphagnum*, structures that are considered analogous to stomata (i.e. pseudostomata) line the equatorial region of the capsule (Boudier 1988). Their function is thought to be purely mechanical, rather than physiological, collapsing upon dehydration to force the release of the operculum. Stomata have been lost repeatedly during the diversification of mosses. In lineages derived from the earliest cladogenic events (e.g. *Takakia* and *Andreaea*), stomata are consistently lacking, suggesting that they originated after the divergence of mosses. Hence, stomata are

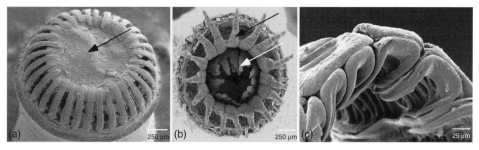

Fig. 4.17. Two fundamental architectures of the peristome. (a) In nemato-dontous peristomes (e.g. *Atrichum undulatum*, courtesy of Neil Bell), the teeth are composed of whole cells. In the Polytrichaceae, the teeth are connected to a thin membrane, the epiphragm (arrow). (b) In arthrodontous peri-stomes, the teeth are composed of remnants of cell walls. The peristome may be composed of two or one ring of teeth. The outer ring is called the exostome (black arrow), the inner one, the endostome (white arrow). Typic-ally, only one wall per cell remains and is attached to the sole remaining wall of the cells facing it (c).

probably not homologous among mosses, hornworts and vascular plants (see Section 1.4.2).

The peristome is a unique attribute of mosses but is not present in all species: this innovation was acquired after the divergence of the Andreaeo-bryopsida (Fig. 4.22) and subsequently lost in many taxa. The peristome consists of one or two concentric rings of teeth exposed following the loss of the operculum (Fig. 4.17). In *Polytrichum* and allied taxa, the teeth are composed of entire elongated cells (Fig. 4.17a) and the peristome is said to be nematodontous. In all other peristomate mosses, the cells contributing to the peristome are partially degraded and the teeth are built almost exclusively from vertical cell walls (Fig. 4.17b, c). The teeth are articulate. They can bend inward or outward, hence the name arthrodontous for this peristomial archi-tecture. As is always the case in bryophytes, movement is accounted for by differential forces acting upon hydration (or dehydration) of cell walls of different thicknesses. Each tooth is composed of two radial columns of plates, an inner and an outer set. As water evaporates, the thicker walls shrink and the collective movement along the columns results in the tooth bending towards that side.

The architecture of the peristome varies and is fundamental to the diagno-sis of major lineages (Fig. 4.18). With the exception of *Mittenia*, cells from three amphithecial layers may contribute to peristome formation: the inner, primary and outer peristomial layer, referred to as the IPL, PPL and OPL, respectively (Fig. 4.18b). Only the outer vertical wall of each IPL remains at

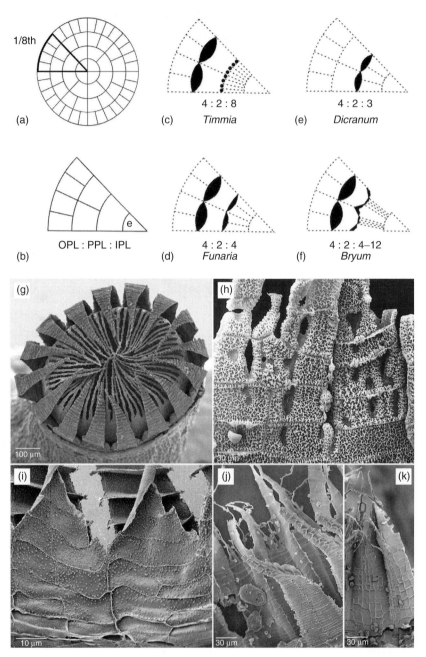

Fig. 4.18. Peristome architecture in mosses. (a) Diagram of a transverse section through the putative peristome-forming region at the apex of an immature moss sporophyte. (b) Detail of an eighth of the section in (a), showing the endothecium, e, and the three innermost amphithecial layers that contribute to peristome formation: outer (OPL), primary (PPL) and inner (IPL) peristomial layer. (c–f) Diagram of an eighth of a *Timmia*-, *Funaria*-,

maturity and only the inner wall of the OPL. In some taxa, one of these layers may be completely disintegrated at maturity. The PPL contributes two walls: the inner one, lying against the IPL wall; and the outer one, attached to the OPL wall. All other walls are degraded (Fig. 4.18c–f). Thus, two concentric rings of teeth may be formed, one composed of the OPL + PPL (the exostome) and the other composed of the PPL + IPL (the endostome; Fig. 4.18c, d, f). Some mosses develop a double peristome, others a simple one, consisting only of either the outer or inner ring and others have lost the peristome completely. The number of cell divisions within the three peristomial layers varies. The PPL is typically composed of 16 cells and the OPL of 32, whereas the IPL has between 16 and 64 cells. Consequently, each PPL cell faces 2 outer cells and 1 to 4 inner cells. The exostome and endostome typically comprise 16 teeth each. The outer teeth are often fused into 8 pairs and more rarely divided into 32 teeth. The complexity of the endostome varies along with the architecture of the IPL (Fig. 4.18c, d, f). The teeth, here called segments to distinguish them from the outer teeth, number 16. Divisions within the IPL resulting in more than 32 cells lead to the formation of additional appendages, which, except in *Timmia* (Fig. 4.18g), are much narrower and smaller than the segments (Fig. 4.18j, k). Endostomial segments and cilia are also often mounted on a basal membrane (Fig. 4.18k). Whereas the divisions within the OPL and PPL are always symmetric, those in the IPL may be conspicuously displaced from the median, resulting in two daughter cells of unequal size (Fig. 4.18e, f). The pattern in number and plane of cell divisions and in the resorption of cell walls defines major peristome types: the diplolepideous opposite (Fig. 4.18d, i) and alternate types (Fig. 4.18f, j), and

Caption for Fig. 4.18. (cont.)
Dicranum- and *Bryum*-type peristome. Black areas identify thickened cell walls composing the peristomes. Dotted lines mark the walls of the IPL, PPL and OPL cells that are resorbed and hence that are not contributing to the peristome. (g) Diplolepideous peristome of *Timmia megapolitana*, showing the 64 filamentous appendages of the endostome. (h) Outer view of the haplolepideous peristome of *Tortula plinthobia*. Each tooth is fenestrate along the vertical walls of the IPL and hence one and a half cells of the IPL face each PPL cell (outer cells in view here). (i) Inner view of the peristome of *Funaria hygrometrica*, showing the four IPL cells composing the two segments, which lie opposite the two exostome teeth. (j) Diplolepideous peristome of *Pseudoscleropodium purum*, showing the keeled endostome segments alternating with the exostome teeth and the cilia between two consecutive segments. (k) Inner view of the diplolepideous peristome of *Mnium thomsonii*, showing the numerous cells composing the IPL. Reproduced from Budke *et al.* (2007) (a–f) and Goffinet *et al.* 2009 (g–k).

the haplolepideous type (Fig. 4.18e, h). In the diplolepideous peristome, the outer tooth is derived from two columns of cells facing each column of PPL cells (Fig. 4.18c, d, f). In the haplolepideous peristome, each tooth bears the remnants of a single column (Fig. 4.18e). Two rings of teeth occur in most diplolepideous peristomes, whereas haplolepideous peristomes consist of only the endostome. The haplolepideous peristome characterizes the Dicranideae (therefore also referred to as haplolepideous mosses), whereas all other arthrodontous peristomes are of the diplolepideous type. Reduction through the loss of one or both rings, of endostomial cilia, or the truncation of the teeth, has occurred many times independently among mosses (Vitt 1981) and in these cases the architecture at maturity can be phylogenetically misleading.

To release the spores from the capsule, mosses either shed an operculum, have sporangia that dehisce along longitudinal lines or irregularly through the breakdown of the capsule wall. *Sphagnum* and most other mosses see their spore mass exposed after the capsule looses an apical lid. The loosening and shedding of the operculum is often triggered by the tensions in the annulus, composed of one or more rings of cells with uneven wall thickenings. In *Takakia*, the capsule dehisces along a spiral line extending nearly to the poles (Fig. 4.2a). In *Andreaea* and *Andreaeobryum*, the lines are vertical converging towards, but ending below, the apex, respectively (Fig. 4.2e, f). Mosses occurring in highly seasonal habitats tend to dehisce along an irregular line or randomly as the capsule wall breaks down or is physically damaged. Peristomes occur only in operculate taxa, a pattern consistent with the function of peristomes in controlling spore release. Only *Tetraplodon paradoxus* bears a peristome, despite the capsule remaining closed at maturity.

Spore dispersal is passive, except in entomophilous Splachnaceae (Box 4.2), and gradual. In *Sphagnum*, spores become airborne following the implosion of the capsule and the ejection of the operculum. During the final stages of sporangium maturation, gases accumulate inside the capsule. Upon dehydration, the capsule wall collapses inward along the equatorial region bearing the pseudostomata. The compression of the gases results in an increase in pressure, forcing the rapid tearing of the capsule along its line of dehiscence and the explosive release of the operculum. In some nematodontous mosses, the capsule mouth is closed by an epiphragm, a thin membrane spanning the mouth (Fig. 4.17a). It bears minute perforations along its circumference, which prevent spores from being released all at once. Similarly, in arthrodontous mosses, the peristome ensures the gradual release of spores. Spreading the dispersal of spores over several days or weeks may increase the likelihood of spores being more widely distributed under different climatic conditions. In some taxa, the peristome overarches the mouth when moist to prevent water

Box 4.2
Fly-mediated spore dispersal in dung mosses

Bryophyte spores are typically dispersed by wind and few may find optimal conditions for their germination and subsequent survival. Targeted dispersal to specific microhabitats would enhance germination rates but requires a vector. Flowering plants often recruit either insects, birds or bats to pick up and deliver pollen grains. Elaborate displays and rewards increase visitations, specificity and effectiveness of delivery. The costs are high, but so are the dividends: fewer pollen grains need to be produced to achieve a given rate of sexual reproduction.

Among seedless embryophytes, only members of the moss family Splachnaceae rely on insects to disperse their spores. Approximately half of the species are coprophilous: they live exclusively on dung or animal remains, which gives the family the common name dung mosses. These substrata are not hostile to other mosses, but the Splachnaceae claim residency before their competitors have a chance to get established. They achieve this by recruiting flies that seek fresh dung and carrion to feed or lay their eggs. To attract the vector, Splachnaceae use deceptive visual and olfactory cues, displayed and emitted by the highly modified sporophyte. The capsules are white, bright yellow, red or deep purple in colour and are often inflated (Box 4.2 Fig. 1). Critical for attracting flies to the sporophyte are the odours mimicking the decaying animal substrata. These volatile compounds are produced by the seta and the apophysis. The chemistry differs between dung and carrion inhabiting species, and the diversity of compounds seems to be inversely correlated to the size of the sporophyte: large, showy sporophytes emit fewer compounds than small, inconspicuous capsules (Marino *et al.* 2009). Taxa with overlapping geographic ranges may partition the habitat or co-exist. In Patagonia, *Tayloria mirabilis* lives on cow dung, whereas *T. dubyi* occurs only on geese dung. *Tetraplodon fuegienus* is restricted to carnivore dung, like its close relatives from the Northern Hemisphere. In Alaska, *Splachnum luteum*, *S. rubrum* and *S. sphaericum* often coexist on the same patch. The three species produce strikingly different odours and their capsules are highly dissimilar in colour and shape. A diversity of dipterans lands on the sporophyte, which they quickly realize is not the substrate they seek. Visitation time is thus short, but long enough for the fly to come into contact with the protruding mass of sticky spores, which cling to its legs and body. When the fly reaches a patch of fresh dung, or carrion, the spores may fall off. Establishment is rapid and soon a dense population of Splachnaceae colonizes this microhabitat. Dispersal to a specific habitat is thus very effective and relies on highly modified sporophytes that function much like flowers. Unlike pollination, insect-mediated spore dispersal requires a single visitation to be successful. Recruiting insects to disperse spores has obvious advantages and may be critical for a species to occupy such patchily distributed substrate as

---

Box 4.2 (cont.)

dung or carrion. However, the costs are high and the constraints severe: the moss must invest many resources into the synthesis of olfactory and visual cues, and success is guaranteed only if the vector is present. Not surprisingly, phylogenetic inferences suggest that reversion to wind dispersal might have occurred once or more during the evolutionary history of the Splachnaceae (Goffinet *et al.* 2004).

Box 4.2 Fig. 1. (a) *Splachnum luteum*: the sterile base of the capsule is expanded to a broad disc, likely designed to display the bright yellow colour and facilitate visitation. (b) *Tayloria dubyi*: a Patagonian endemic restricted to goose dung on *Sphagnum* hummocks in bogs. (c) Simple trapping experiments allow researchers to study the diversity of insects recruited by the mosses. See plate section for colour version.

---

from entering the capsule and triggering the premature germination of spores. The endostome often lacks conspicuous hygroscopic movement in double peristomate mosses. Such attributes may be critical for taxa with pendent capsules to avoid loosing the entire spore mass at once.

The peristome is not the sole structure designed to control spore dispersal. Modifications of the wall or the internal architecture of the capsule may also contribute to regulating the release of spores. Through contraction and expansion in response to changes in atmospheric moisture, the capsule may constrict at the mouth or, more commonly, in the region immediately below. Such changes are, as in other cases of movement, accounted for by uneven degrees of cell wall thickness. This is best seen in ribbed capsules, wherein the ribs are defined by vertical rows of thick-walled cells alternating with bands of thin-walled cells. In some cases, the shrinking is axial, due to thin longitudinal and thick horizontal walls. This movement may promote the exposure of the spore mass upon drying. In *Scouleria aquatica*, a semi-aquatic species, the operculum is retained on a persistent columella. Spores are dispersed following the vertical contraction of the capsule when exposed to dry air. When moist, the capsule expands and is closed by the lid.

Spores are unicellular. In rare cases the spore, undergoes divisions prior to dispersal. The wall of the spore is composed of sporopollenin, a compound that confers high resistance to mechanical and physiological stress. The outer layer is often patterned and the ornamentation may provide critical information for distinguishing species (Fig. 4.19).

For the completion of its development, the sporophyte depends exclusively on the maternal gametophyte to acquire nutrients and water. Throughout much of its growth, the presumptive sporangial region of the sporophyte is covered by a hood. Following fertilization, the archegonium acquires

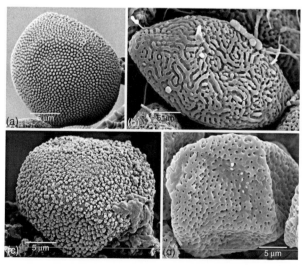

Fig. 4.19. Diversity in spore ornamentation. (a) Papillose (*Funaria hygrometrica*). (b) Lirellate (*Rhachitheciopsis tisserantii*). (c) Granulose (*Schlotheimia tecta*). (d) Pitted (*Rhachithecium papillosa*).

a protective function for the developing embryo. The venter and the cauline tissue immediately below form an epigonium, or sac, enclosing the young sporophyte. The epigonium is of determinate size. The pressure resulting from the growth of the sporophyte causes the protective sac to tear. The basal section remains around the base of the seta and the remainder forms a hood, or calyptra, covering part or all of the sporangium. Removal of the calyptra prior to sporogenesis results in aberrations and sporophyte abortions, although it must eventually be shed to allow for capsule dehiscence and spore dispersal. The mechanisms by which the calyptra controls sporophyte development are not fully understood and may be physiological or solely physical. The calyptra offers critical taxonomic characters, diagnostic of species or lineages of higher rank. It varies in size (covering only the operculum or enclosing the entire capsule), in outline (from conical to bell-shaped or long cylindrical), in basal lobing (from entire to broadly or finely incised), in ornamentation (from glabrous to pubescent), in surface relief (from smooth to ridged) and in surface roughness (from smooth to papillose; Fig. 4.20). The calyptra is rarely persistent. Typically, it is blown off by wind. This is facilitated by loosening the fit of the calyptra on the sporangium, either by elevating it via the growth of a long pointed rostrum on the operculum or via curving of the capsule. Cucullate calyptrae bear a longitudinal slit and are typically associated with asymmetric capsules or oblique rostra (Fig. 4.20). Mitrate calyptrae lack such tearing.

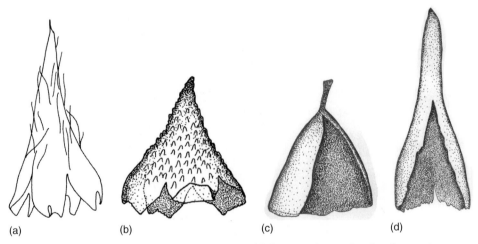

(a)   (b)   (c)   (d)

Fig. 4.20. Variation in calyptra shape. (a) Long mitrate (and pubescent or hairy) and lobed at base. (b) Short, entire and mitrate. (c) Short cucullate. (d) Long cucullate.

4.1.3 Asexual reproduction

In all seedless plants, including mosses and other bryophytes, the flagellated sperm cells must swim to the egg. In species with unisexual gametophytes, a few centimetres separating the sexes may suffice to severely limit the likelihood of fertilization. When male and female gametophytes grow in close proximity, the need for a direct aquatic connection between the two sex organs may be lessened either by recruiting tiny arthropods as sperm vectors (Cronberg *et al.* 2006a), or by projecting sperm cells beyond the immediate vicinity of the male plant using raindrops falling on antheridia clustered in terminal splash cups. Perhaps as many as 50% of mosses are dioicous and, for these, sexual reproduction may be severely constrained. In extreme cases, only male or female plants occur on a continent and, hence, sexual reproduction is completely lacking. Many mosses, like liverworts, rely on asexual means to maintain populations at the local and maybe even regional level. Asexual reproduction is, however, not confined to species with unisexual gametophytes and this strategy may be adopted by some monoicous species for rapid expansion following initial establishment.

Propagules developed for clonal reproduction are diverse. They vary in their shape, size, mode of abscission, longevity and origin on the plant, and may occur on protonema, rhizoids, stems and leaves (Fig. 4.21). A classification of propagules has been proposed by Newton and Mishler (1994) and Laaka-Lindberg *et al.* (2003). A single species may, at least in culture, produce two

Fig. 4.21. Diversity of vegetative propagules in mosses. (a) Multicellular leaf gemmae in *Syrrhopodon texanus* (photo P. Davison). (b) Rhizoidal tuber in *Bryum rubens*. (c) Caducous leaves in *Campylopus pyriformis* (photo P. Degroot). See plate section for colour version.

or three diaspore types (Duckett & Ligrone 1992). *In vitro*, mosses can be propagated from fragments relatively easily (Duckett *et al*. 2004), which facilitates the *ex situ* conservation of rare and endangered species (Section 10.3.3). Ground-dwelling mosses may be prone to breakage caused by animals and new individuals may be established from detached leafy shoots. A recent report even suggests that leafy shoots may remain viable after passing through the digestive tract of bats (Parsons *et al*. 2007). In nature, programmed fragmentation is rare and specialization involves caducous leaf-tips or leaves (e.g. *Campylopus pyriformis*; Fig. 4.21c), plates of leaf blades (e.g. *Zygodon fragilis*), or branches with small leaves (e.g. *Orthodicranum flagellare*). In *Pohlia*, highly reduced branches called bulbils bear only a few leaf primordia. Cauline or foliar gemmae consist generally of multiple cells forming either short filaments or club-shaped structures (e.g. *Syrrhopodon*; Fig. 4.21a). Some species rely on rhizoidal tubers, perennating structures buried in the soil, to maintain the population in seasonal habitats (e.g. *Bryum rubens*; Fig. 4.21b). Unlike gemmae, tubers lack a distinct abscission cell and their development seems to be triggered by drought. Tubers occur in many lineages of mosses, but are conspicuously lacking among pleurocarpous mosses.

4.1.4 Symbiotic associations

A broad array of fungi, distributed across all three lineages (i.e. Glomeromycota, Ascomycota and Basidiomycota) inhabit the leafy shoots and in some cases even the sporophyte of mosses (Davey & Currah 2006). The nature of the interaction varies and is revealed by the distribution of the fungus on the plant and the responses developed by the moss. Healthy tissues may harbour commensals, pathogens or parasites and suffer no, little or severe damage, even death. Endophytic mutualists may lead to physiological changes, increasing tolerance of the host to environmental stress and broadening the ecological amplitude or the resistance to pathogens of the host. Although mosses harbour fungi known to establish mycorrhizal associations with other plants (Zhang & Guo 2007), evidence for the mutualistic nature of the interaction and thus for the transfer of water and nutrients, is lacking (Davey & Currah 2006).

Mosses also commonly associate with cyanobacteria that grow as epiphytes between or on leaves and stems, or in the case of *Sphagnum*, even inside the hyalocysts (Solheim & Zielke 2002). Internal canals or external vesicles harbouring the prokaryote, as found in liverworts and hornworts, are lacking. Establishing a partnership with cyanobacteria offers mosses a valuable source of nitrogen, the growth-limiting macronutrient for most plants. At a broader scale, such associations may result in a substantial input of fixed nitrogen to the ecosystem (see Section 2.1). The cyanobacterium benefits from the protection of the moss (either simply in the form of shelter from extreme abiotic conditions or through chemical optimization of the microenviroment by the moss) as well as from sugars leaked or potentially secreted by the host (Adams & Duggan 2008). Occasionally, green algae may occur inside the laminal cells of mosses (Reese 1981), but such endophytes certainly carry no benefits to the host. Mosses, like most other land plants, are also frequently colonized by heterotrophic prokaryotes, particularly methylobacteria, which consume methanol and other small organic molecules emitted by the host. Although the moss plant may not gain any nutrients from its epiphyte, these bacteria produce cytokinins and other phytohormones that promote the growth of the plant (Hornschuh *et al.* 2002). The hyalocysts of submerged *Sphagnum* in peatlands further harbour methanotrophic bacteria that feed on methane released by methane-producing bacteria that also inhabit these cells. The methanotrophic bacteria oxidize the methane into CO_2, providing the moss with a significant source of carbon (Raghoebarsing *et al.* 2005). Evidently, moss gametophytes compose complex miniature ecosystems within which nutrients are extensively recycled.

4.2 Classification and macroevolution

Most mosses are readily assigned to taxonomic classes based on morphological characters. The Takakiopsida are defined by deeply incised leaves and a spiral line of sporangial dehiscence. Dimorphic cells, with photosynthetic and hyaline cells forming a unistratose lamina, branches arranged in fascicles and an operculate capsule elevated on a pseudopodium diagnose the Sphagnospida. The Andreaeopsida and Andreaeobryopsida occur on rocks and bear a valvate capsule. Nematodontous peristomes identify the Polytrichopsida (with 16 to 64 teeth) and Tetraphidopsida (with only 4 teeth). Finally the Bryopsida, the most diverse class of mosses, comprises all species with an arthrodontous peristome. Intuitively, the Bryopsida have been considered the most derived class of mosses, sharing a common ancestor with other peristomate mosses (Vitt 1984). The remaining classes arose earlier in the evolutionary history of mosses, with the origin of the Sphagnopsida predating the divergence of the Andreaeopsida and Andreaeobryopsida (Newton *et al.* 2000). Inferences from DNA sequence data corroborate this broad evolutionary scenario. Several nodes, however, still lack strong support (Fig. 4.22). The affinities of the Takakiopsida remain ambiguous (sister to all mosses or only to the Sphagnopsida). Similarly, the shared unique ancestry of

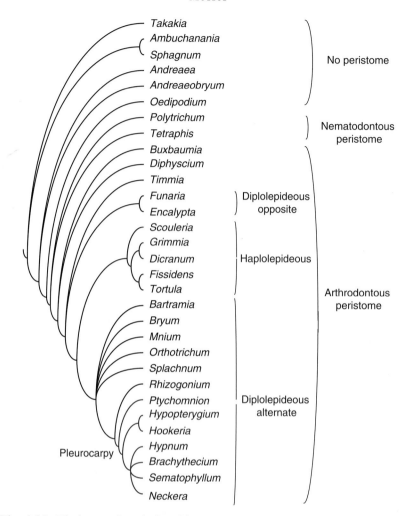

Fig. 4.22. Phylogenetic relationships among major lineages of mosses.

the Andreaeopsida and Andreaeobryopsida or of the nematodontous mosses is still dubious. The difficulty in reconstructing the phylogeny of mosses may be explained by the dramatic morphological divergence and, hence, lack of conspicuous homology among the taxa, or a rapid succession of early clado-genic events that was followed by the accrual of significant homoplasy over the long evolutionary history of mosses. Fossils offer little valuable chrono-logical information since most preserved specimens date from more recent periods than the estimated time of divergence. Dating times of divergences based on calibrated molecular phylogenies (Newton *et al.* 2007) suggests that mosses arose about 380 mya, at a time when forests appeared. The radiation of major lineages (i.e. classes) was completed nearly a 170 my later with the origin of the arthrodontous mosses (Newton *et al.* 2007). These estimates do not support the hypothesis of a rapid early radiation, suggesting that the weaknesses in the backbone phylogeny of mosses results from severe homoplasy or insufficient divergence among the lineages in the loci surveyed so far. Resolving the sequence of divergence events requires extensive sampling of genetic characters and also the exploration of other sources of characters, particularly ontogenetic (Duckett *et al.* 2004) and even biochemical or physiological traits.

Molecular characters have, however, proven critical for revealing the phylogenetic significance of genera such as *Oedipodium* and *Timmia*. The gametophyte of the former resembles that of acrocarpous arthrodontous mosses, yet inferences from various genetic loci suggest a much older origin for *Oedipodium* (290 mya, Newton *et al.* 2007), which shares a common ancestor with all peristomate mosses (Fig. 4.22). *Oedipodium*, however, lacks a peristome; whether this absence is truly a plesiomorphy or results from a loss is not clear. *Timmia*, with its unique endostome of 64 appendages, is reconstructed to have arisen early in the evolutionary history of truly arthodontous mosses rather than derived within the clade of mosses with alternate peristomes. The developmental pattern of its endostome is congruent with such view (Budke *et al.* 2007). Its exact affinities, however, as sister either to all arthrodontous mosses (except *Buxbaumia* and *Diphyscium*) or only to the diplolepideous mosses with opposite peristomes, remain unresolved. The contribution of sequence data is similarly significant for resolving the affinities of taxa with reduced morphologies. The classification of the arthrodontous mosses is drawn primarily on the architecture of the peristome. Within such a system, the affinities of mosses lacking a peristome can be ambiguous. For example, the genus *Ephemerum*, which is characterized by small gametophytes and sessile sporophytes lacking a differentiated line of dehiscence, has traditionally been accommodated within the Funariales (diplolepideous opposite peristome), on the basis of the smooth laminal cells. Yet, inferences from nucleotide substitutions reveal that it likely arose from an ancestor shared with members of the Pottiaceae (haplolepideous peristome) through severe reduction (Goffinet & Cox 2000). Patterns in spore development also support this hypothesis (Duckett *et al.* 2004).

The ambiguity in the phylogenetic tree precludes reconstructing the origin of major innovations and hence the timing and the polarity of some character transformations remain obscure. In particular, the evolutionary and ontogenetic relationships of the various arthrodontous peristome architectures remain unresolved. Budke *et al.* (2007) have recently shown that *Timmia* undergoes a sequence of divisions in the amphithecium that is similar to that of the Funariaceae, but, considering the ambiguity of the position of *Timmia*, it is unclear whether the developmental pattern is plesio- or apomorphic in mosses. The polarity and relative timing of other character transformations, however, is rather unambiguous. The ancestral mosses lacked a costa in their leaves, stomata in their capsule wall, an operculum and a peristome. All of these features occur in peristomate mosses and may have been acquired rapidly, so as to appear linked in the phylogeny.

The classification of mosses continues to be refined based on phylogenetic inferences aimed at reflecting the evolutionary history of mosses rather than mere patterns in morphological similarity. Most profoundly affected are the concepts of familial delineation within the pleurocarpous mosses (i.e. Hypnales and Hookeriales, in a broad sense), which account for nearly 50% of moss diversity. A rapid diversification following the transfer of sex organs to short lateral branches, and thereby the ability to maintain uninterrupted vegetative growth, combined with almost immediate multiple independent adaptive radiations exploiting an increase in habitat diversity correlated to the rise of angiosperms (Newton *et al.* 2007), may explain the difficulty in reconstructing the relationships among pleurocarpous mosses.

4.3 Biogeography and ecology

Like liverworts, mosses occur on all continents. They grow on a wide range of substrata, including soil, rock, bark, rotting wood, dung, animal carcasses and leaf cuticles. The only ecosystem that mosses (and bryophytes in general) have failed to colonize are oceans. No moss lives permanently in the ocean, although the aquatic moss *Fontinalis dalecarlica* is able to grow in the northern Baltic Sea owing to its low salinity (Bates 2009). A few species, such as the seashore moss *Schistidium maritimum* (Grimmiaceae), are also found in the spray zone in coastal habitats.

Traditionally, mosses have been regarded as fairly faithful indicators of particular habitat conditions (Bates 2009). The success of mosses in colonizing almost all available habitats is due to a high turnover of specialist species rather than the wide occurrence of 'generalist' species. This specificity already appears at a high taxonomic level, meaning that the descendents of a common ancestor tend to occupy similar ecological niches, a phenomenon known as niche conservatism. For example, ombrotrophic bogs, which receive water and nutrients only from precipitation, are always dominated by *Sphagnum* (Vitt & Wieder 2009). Rich fens, which are fed by both precipitation and groundwater and are characterized by a higher pH, are dominated by a suite of species belonging to the two pleurocarpous moss families Calliergonaceae and Amblystegiaceae, as if the supposedly adaptive traits of their common ancestor constrained the daughter species to remain in the same environment (Hedenäs 2001). Similarly, all of the 50 species of *Andreaea* are restricted to hard rocks with a low calcium concentration (Heegaard 1997).

In contrast to the highly specific angiosperm flora that typically characterizes harsh habitats (e.g. salt marshes or metal-contaminated soils), mosses tend to exhibit a more generalist strategy. For example, the moss species composition along a salinity gradient does not show the zonation pattern that characterizes the vascular plant vegetation in salt marshes (Zechmeister 2005). Adaptations to extreme habitats occur but are by no means the rule in mosses. Metal-contamination in soils may be revealed by the presence of *Scopelophila cataractae*, which is strictly restricted to this substratum, but does not exclude other moss species from surrounding, non-polluted soils from becoming established. Similarly, several moss species common in non-saline environments, such as the forest moss *Mnium hornum* or the freshwater marsh species *Campylium stellatum*, display a considerable salinity tolerance (Garbary *et al.* 2008). Mosses may thus exhibit a much broader, albeit hidden, ecological amplitude than is revealed by their typical realized niche. Such potential would eliminate the need for specialization (see Chapter 9). Alternatively, specialization may be purely physiological rather than morphological and thus not be apparent to the taxonomist.

5

Hornworts

Hornworts (Division Anthocerotophyta) compose the least diverse lineage of bryophytes. Their name refers to the horn-like sporophyte that lacks a seta and dehisces along one or two vertical lines (Fig. 5.1a). They further differ from mosses by the lack of leaves. Their vegetative body is indeed thalloid and rosette or ribbon-shaped and thereby resembles that of some liverworts. Also like liverworts, hornworts lack conspicuous and constant diagnostic gametophytic features and assigning a thallus to one lineage or another relies on microscopic characters. Although morphologically rather simple in architecture, hornworts may be physiologically elaborate organisms. Many species establish and maintain intimate symbiotic associations with endophytic nitrogen-fixing cyanobacteria and endomycorrhizal fungi. Furthermore, hornworts have carbon-concentrating mechanisms lacking in other bryophytes. Phylogenetically, hornworts appear more derived than mosses and liverworts, too: recent studies suggest that they alone share a unique common ancestor with vascular plants. Although easily overlooked in the field because of their often thin and when dry, dark thallus and neglected by biologists for their low global diversity, hornworts are drawing increasing attention for their significance in the evolution of land plants and their partnerships with endosymbionts.

5.1 Structure and development

5.1.1 Gametophyte

The vegetative body of hornworts is a flattened bilaterally symmetric thallus (Fig. 5.1). Although the thallus lacks organized external appendages, the margin can be deeply incised, giving the appearance of irregular leaves emanating from a thickened midrib in some taxa (Fig. 5.1b). In some cases,

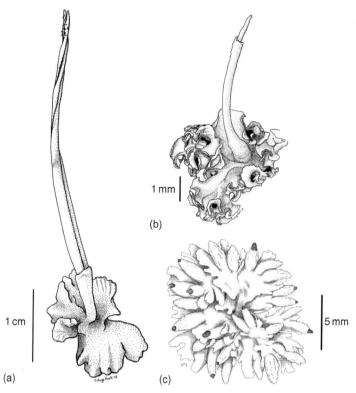

Fig. 5.1. Variation of thallus morphology in hornworts. (a) Typical thallus in rosette in *Phaeoceros*. Note the horn-like sporophyte that dehisces along one or two vertical lines. (b) Deeply incised thallus margins in *Dendroceros*, giving the appearance of irregular leaves emanating from a thickened midrib. (c) *Notothylas orbicularis*, a vigorous plant with mature and immature sporophytes, which are reduced in size owing to the terminate development of the basal meristem (reproduced from Schuster 1966–1992 with permission of the Field Museum of Natural History, Chicago).

the margin may be recurved downwards, forming narrow capillary spaces that may function to hold excess water. The growth of the thallus results from the activity of the single-celled apical meristem, which is located in thallus notches and covered by mucilage secreted by epidermal cells (Fig. 5.2). The geometric shape of the apical cell (Fig. 5.3) determines the growth form of the gametophyte, which can appear as either a rosette (Fig. 1.3) or a ribbon (Fig. 5.4), and also to a lesser extent of the thickness of the thallus (i.e. tapering from the centre to abruptly thinned from the midrib outwards). The thallus lacks conspicuous internal differentiation (Fig. 5.14c), except for the presence of cavities borne through schizogeny, which are essential for symbiotic cyanobacteria. Most parenchyma cells may synthesize, store and secrete mucilage, a carbohydrate that may be essential for water retention.

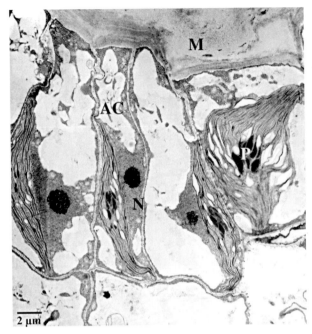

Fig. 5.2. Longitudinal section of growing thallus notch overarched by mucilage, M, and showing rectangular apical cells, AC, with well-developed
nucleus N and chloroplast, P (reproduced from Renzaglia *et al.* 2009 with
permission of Cambridge University Press).

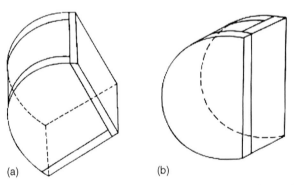

Fig. 5.3. Wedge-shaped segmenting along four cutting faces (a) and hemi-
scoid apical cells cutting along three faces (b) resulting in a rosette (Fig. 1.3)
or ribbon-like (Fig. 5.4) growth form (reproduced from Renzaglia *et al.* 2009
with permission of Cambridge University Press).

The upper surface of the thallus harbours the sex organs and, in some cases,
small dissected flaps or lamellae (Fig. 5.5). Their function is unknown. The
lower surface of the thallus bears smooth unicellular rhizoids that are typically
unbranched, except at their tips in *Megaceros*.

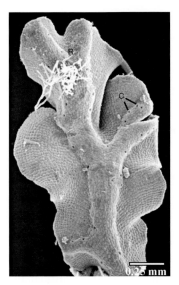

Fig. 5.4. Ribbon-shaped growth form of *Dendroceros* showing swollen central midrib and lateral wings. Note the presence of pore-like mucilage clefts, C, and unicellular rhizoids, R (reproduced from Renzaglia *et al.* 2009 with permission of Cambridge University Press).

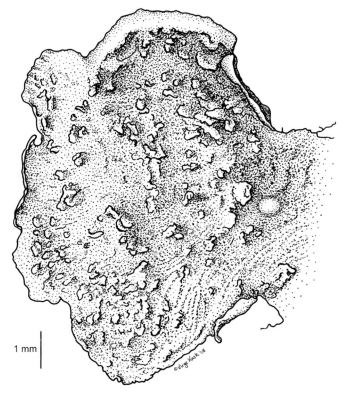

Fig. 5.5. Small dissected flaps on the dorsal thallus in *Phaeomegaceros squamuligerus*.

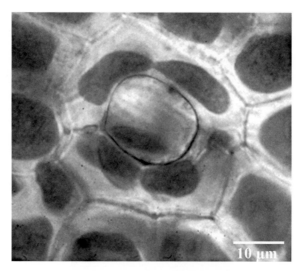

Fig. 5.6. Surface view of mucilage cleft in ventral thallus epidermis (reproduced from Renzaglia *et al.* 2009 with permission of Cambridge University Press).

Upon closer examination, the lower (and rarely also the upper) epidermis reveals small pores defined by two kidney-shaped cells, that very much resemble the stomatal guard cells, as seen in the sporophyte (Fig. 5.6). Like the guard cells in the sporangial wall, these cells lack the ability to open and close the pore; in fact the pore, once formed, remains open. The pore leads to a small chamber that may hold globose colonies of *Nostoc* embedded in mucilage. The 'stomata' are hence called mucilage clefts. Stomata are a necessity only in plant bodies covered by a cuticle, which would limit free gas exchange over the surface of the plant. Hornworts may have a cuticle, but probably not a thick and effective one considering that water is still absorbed over the whole surface of the thallus. Furthermore, the function of the stomata complex relies on the pore opening into an air chamber and a lax parenchyma to allow for the diffusion of carbon dioxide and oxygen. Schizogenous cavities are frequent in hornworts and some are formed above the epidermal pore. They are, however, filled with mucilage and, hence, are inadequate for gas exchange. Whether mucilage clefts are homologous to stomata, as argued by Schuster (1984c), has been debated, and rejected by some researchers, in part on the basis of the functional (and therefore probable ultrastructural) differences (Renzaglia *et al.* 2007). The guard cells of early polysporangiophytes also resemble those of modern taxa but in some species the stoma is not connected to a substomatal chamber (Edwards *et al.* 1998), raising the possibility that stomata did not have the same function as in extant tracheophytes. The possibility remains that mucilage clefts in the gametophyte of hornworts are stomata that have been co-opted from the

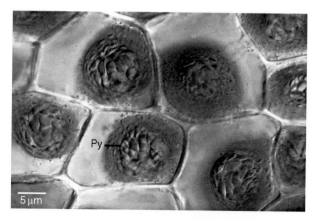

Fig. 5.7. Upper epidermal thallus cells of *Megaceros cf. vincentianus*, each containing a single, large chloroplast with abundant starch and pyrenoid, Py (reproduced from Renzaglia *et al.* 2009 with permission of Cambridge University Press).

sporophyte to fulfil a new function, in which case their evolutionary history is connected. Indeed, the pore serves not for gas exchange but as the entry point for filamentous *Nostoc*, which will form the endophytic colonies characteristic of all hornworts (see Section 5.1.4).

The photosynthetic cells, including the epidermal cells of the gametophyte, typically possess a large solitary chloroplast. Numerous, small plastids characterize only a few taxa, such as *Megaceros*. In either case, a single chloroplast occurs in the cell at the time of cell division. Rubisco, the enzyme responsible for carbon fixation and, hence, essential to photosynthesis, is often concentrated in a discrete region of the chloroplast forming what is called the pyrenoid (Fig. 5.7). This feature is unique among land plants but known from algae, including the Charophycean ancestor to embryophytes.

Most hornworts are monoicous, with individual thalli producing both types of gametangia. Dioicy is less common and sexual dimorphism between unisexual thalli weakly pronounced. Gametangia are produced at the dorsal thallus midline (Fig. 5.8). Archegonia are shaped like a vase, with a slightly swollen venter holding the single egg and a short neck, capped by two to four cover cells; at maturity the central canal cells of the neck disintegrate and the cover cells are shed (Fig. 5.9). The female sex organs develop from an epidermal-initial and remain mostly sunken at maturity with only the apex of the neck protruding from the surface (Fig. 5.14b). Sperm cells are protected by a unistratose jacket subtended by a short stalk. The development of antheridia is similar to that seen in other bryophytes, except that the initial is borne immediately below the epidermis. Male gametangia always develop within a chamber and typically occur in aggregates, with all antheridia derived

Fig. 5.8. Antheridia and mature sporophyte in *Phaeoceros laevis* thriving on steep, wet rocky slopes of humid laurel forests in La Palma, Canary Islands A: antheridia; C: capsule; I: involucre. Note the yellowish apex of the capsule from which yellow spores, S, are released, while new spores are still being produced at the green base of the capsule (photo A. Vanderpoorten). See plate section for colour version.

from a single initial (Fig. 5.10). The epidermal cell above the antheridial initial gives rise to the two-layered roof of the antheridial chamber. A schizogenous cavity is formed following the separation of the two initials early in gametangial ontogeny. At maturity, the location of the endogenous antheridia is revealed by the convex ceiling of the chamber on the thallus surface.

5.1.2 Sporophyte

The plane of the first division of the zygote is longitudinal, unlike in other bryophytes where it is transverse. Early in its development, the embryo is

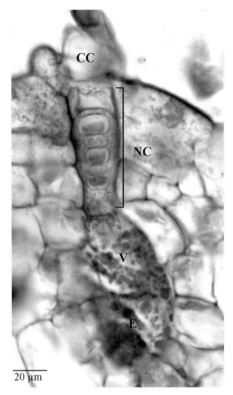

Fig. 5.9. Longitudinal section of an archegonium in *Phaeoceros laevis* with two cover cells (CC), six neck canal cells (NC), a ventral canal cell (V) and an egg cell (E) (reproduced from Renzaglia *et al.* 2009 with permission of Cambridge University Press).

divided into a basal region that produces the somewhat enlarged foot and an upper region that develops into a capsule that grows acropetally from a basal meristem. A stalk subtending the sporangium, like the seta in mosses and liverworts, is always lacking.

The peripheral cells of the foot form a haustorium that penetrates the surrounding gametophytic tissue to form the placenta, characterized by a large exchange surface between the two generations, favouring the efficient nourishment of the sporophyte (Fig. 5.11). A distinctive feature of the placenta of some hornwort species is the occurrence of abundant protein crystals, which likely derive from gametophytic cells and may be a source of amino acids for the developing sporophyte, and the presence of wall ingrowths only in the gametophyte.

The sporangium grows from the activity of a basal meristem. The cells deriving from the meristem follow two distinct developmental patterns. The axial cylinder composes the endothecium, which will form the columella.

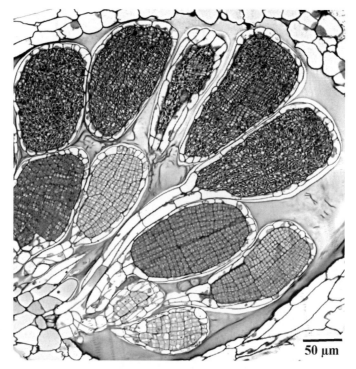

Fig. 5.10. Section of antheridial chamber with 11 antheridia at different stages of development (reproduced from Renzaglia *et al.* 2009 with permission of Cambridge University Press).

The outer cylinder, or amphithecium, develops into the epidermis, the assimilative layer and, unlike in other bryophytes (except *Sphagnum* and *Andreaea*), also the sporogenous tissue. The basal meristem remains active throughout the life of the sporophyte, except in *Notothylas*. In the latter, the basal meristem functions for a limited period. The sporophyte, hence, remains small and is frequently retained within the protective tissue of the gametophyte (Fig. 5.1c). The continuous growth of the sporophyte, unique among land plants, results in a continuous and, hence, non-synchronized production of spores and an acropetal gradient of their maturation. Consequently, the different stages of spore development beginning with meiosis near the base of the sporangium, can thus be observed within a single sporophyte. In mosses and liverworts, all sporocytes undergo meiosis more or less synchronously and at the time of dispersal all spores have completed their maturation.

When young, the capsule is protected by a multilayered involucre, developed from the gametophytic tissue enclosing the archegonium (Fig. 5.11). The involucre typically ceases to grow when the first meiotic divisions occur in the sporangium. Through continuous basal growth, the sporophyte soon

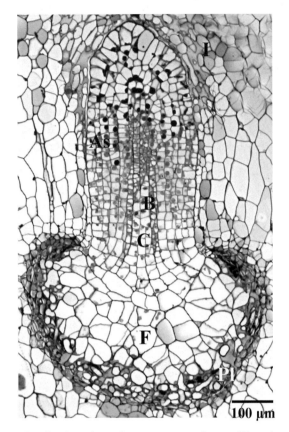

Fig. 5.11. Longitudinal section of young sporophyte still enclosed within the protective gametophytic involucre, I. The prominent foot, F, consists of large ventral cells and smaller peripheral cells that interdigitate with gametophytic cells to form the placenta, Pl. A basal meristem, B, has begun to produce a columella, C, and assimilative tissue, As (reproduced from Renzaglia *et al.* 2009 with permission of Cambridge University Press).

ruptures the involucre near its apex and emerges from the gametophytic cylinder (Figs. 5.8, 5.14a). By contrast, in *Notothylas*, the sporophyte remains short, emerges only slightly from the involucre and, hence, completes its development while enclosed by the gametophytic protective tissue. Stomata occur in many, but not all species and open to a substomatal chamber when the epidermis is exposed above the involucre. The stomata are morphologically similar to those of vascular plants and defined by two guard cells (Fig. 5.12). These cells seem to lack an ability to open and close (Lucas & Renzaglia 2002) and rather than control water and gas exchange in relationship to photosynthesis, stomata may be essential for the dehydration and dehiscence of the sporophyte (see Section 1.4.2). The spore mass is indeed embedded

Fig. 5.12. Stoma in sporophyte epidermis of *Leiosporoceros* showing two
guard cells surrounding a median pore (reproduced from Renzaglia *et al.*
2009 with permission of Cambridge University Press).

in mucilage that protects the developing spores by keeping the tissue hydrated;
once the spores are mature, the mucilage must dry out for spores to be shed
freely. Spores vary among species in size, colour, wall thickness and longevity.
Yellow and brown spores with thick walls tend to be long-lived (up to 21 yrs in
herbarium packets). Longevity may be accounted for by the oils filling the
spores, with the oil functioning as nutrient storage and protection against
desiccation. Green spores are restricted to tropical and subtropical genera.
Their colour comes from the large chloroplast that is visible through the thin
translucent wall. These spores lack oils and are short-lived. Features of the
spore, especially colour and wall ornamentation, as well as of the pseudo-
elaters, are central to species delineation in hornworts (Fig. 5.13).

Before sporogenesis, the cells of the sporogenous region undergo a division
that will yield a spore mother cell (or sporocyte) and a pseudo-elaterocyte.
Pseudo-elaters consist of one or more elongate diploid cells. Multicellular

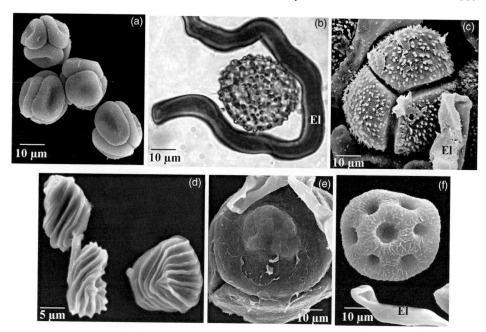

Fig. 5.13. Variation in spore and pseudo-elater (El) shape in hornworts. (a) Tetrad of smooth spores in bilateral arrangement in *Leiosporoceros dussii*. (b) Papillose spore and thick-walled pseudo-elater in *Folioceros appendiculatus*. (c) Tetrad of spinose spores surrounded by short, smooth pseudo-elaters in *Phaeoceros carolinianus*. (d) Spores with striate-canaliculate ornamentation in *Hattorioceros striatisporus*. (e) Tetrad of spores with prominent mammilla in *Phymatoceros phymatodes*. (f) Spore with vermiculate surface with six depressions around a larger, central one in *Phaeomegaceros fimbriatus* (reproduced from Renzaglia *et al.* 2009 with permission of Cambridge University Press).

pseudo-elaters are filamentous. The cell walls may be thin, evenly or spirally thickened (Fig. 5.13). Spores and pseudo-elaters are typically arranged in alternating layers inside the sporangium. Spore mother cells enter meiosis immediately following the isolation from the pseudo-elater mother cell. The latter typically undergoes repeated mitotic divisions; as a result, pseudo-elaters outnumber the spores in the mature sporangium. Pseudo-elaters are very similar to the elaters of liverworts. Both trace their cell lines to a shared ancestry with spore mother cells, but the pattern of cell divisions, including the interpolation of mitotic divisions prior to cytological differentiation, differs and is interpreted by some as indicative of analogy rather than homology of these structures.

Sporophyte dehiscence occurs near the apex by splitting along one or two longitudinal lines with the valves sometimes remaining attached apically (Figs. 5.1, 5.8). Separation and dispersal of the spores is often facilitated by the twisting of the capsule wall and pseudo-elaters that disrupts the spore mass upon drying. Spores maintain their tetrahedral arrangement until nearly mature and, hence, often exhibit a conspicuous trilete mark that reveals the pole of the spore touching the three other spores within a tetrad. Except in *Dendroceros*, germination of the spore occurs after dispersal and is exosporic, with the germ tube protruding from the spore. The sporeling consists of a globose mass from which rhizoids emerge. Soon, an apical cell is differentiated and a thallus develops.

5.1.3 Asexual reproduction

Most hornworts develop both sex organs on a single thallus and sexual reproduction is frequent for these taxa. Highly specialized means of asexual reproduction are uncommon. Some annual species develop perennating tubers, either along the thallus margin or on the lower surface. Others develop gemmae on the upper surface and its edges. The simplest form of vegetative reproduction occurs through older parts of the thallus dying off, while the younger parts, now disconnected, continue to grow. In dioicous species and particularly when sexes are not sympatric, survival of the populations depends entirely on the species' ability to generate new individuals clonally. One example is that of *Megaceros aenigmaticus*, a species known only from the Southern Appalachians in the eastern United States. Male and female plants occur in different watersheds and are thus geographically isolated. Locally the species is abundant, reproducing seemingly exclusively through fragmentation along the thallus margin (Renzaglia & McFarland 1999).

5.1.4 Symbiotic associations

Hornworts are well known for their endosymbiotic *Nostoc* colonies. *Nostoc* is a photosynthetic, filamentous, soil-dwelling cyanobacterium that can develop symbiotic associations with fungi, liverworts and even cycads. *Nostoc* filaments penetrate the hornwort thallus through clefts on the lower epidermis. Inside the gametophyte, it forms spherical colonies that are soon surrounded by mucilage secreted by the parenchyma cells. *Nostoc* is recruited for its ability to fix atmospheric nitrogen, which is highly abundant but not directly usable by most organisms. In *Nostoc*, nitrogen fixation takes place in specialized non-photosynthetic cells called heterocysts. Hornworts tap into this supply of a crucial macronutrient through haustoria-like cell extensions that penetrate the colony. *Nostoc* in return receives organic carbon from the hornwort

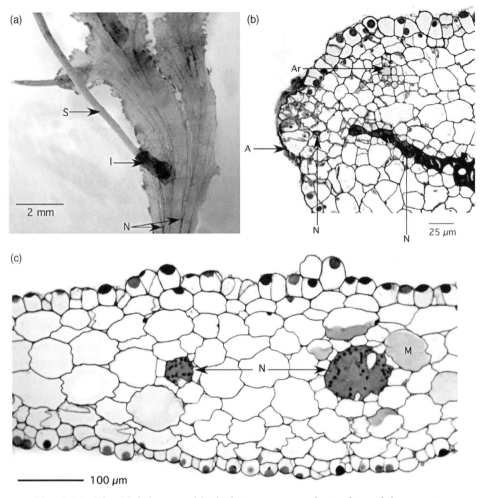

Fig. 5.14. The N-fixing symbiosis between cyanobacteria and hornworts. Example of the *Nostoc-Leiosporoceros* association (reproduced from Villareal & Renzaglia 2006 with permission of *American Journal of Botany*). (a) Dorsal view of the thallus showing bifurcating strands of *Nostoc* parallel to the main axis of the thallus (arrows). Note the mature sporophyte, S, with a basal involucre I. (b) and (c) Longitudinal (b) and transverse (c) section of the thallus with *Nostoc* canals (arrows) originating behind the apical cell A and scattered mucilage cells (M). Note the presence of an immersed archegonia Ar within the thallus.

(Duckett *et al.* 1977, Rodgers & Stewart 1977). In *Leiosporoceros, Nostoc* must be acquired at the sporeling stage, since clefts are lacking in mature tissues. The *Nostoc* colonies typically form a globose mass inside the cavity, except in *Leiosporoceros*, wherein the cyanobacterium elongates within long more or less axial canals (Fig. 5.14). In symbiotic colonies, heterocysts compose nearly 50% of the *Nostoc* cells, as also occurs in other, well-documented cases of symbioses involving cyanobacteria, e.g. in the fern *Azolla* and cycads. Some hornworts also establish mycorrhizae-like associations with aseptate fungi of the Glomales (Schüßler 2000). Whether such fungal partnerships are widespread among species and obligatory for the hornwort involved needs further study.

5.2 Classification and macroevolution

Today, approximately 300 hornwort species, 14 genera and five families, are recognized (Duff *et al.* 2007). Characters used to distinguish taxa include the presence of mucilaginous cavities, stomata and pyrenoids; the number of chloroplasts per cell; the number of antheridia per chamber; the ornamentation of the spore wall; and the shape of the pseudo-elaters. The chloroplasts of hornworts exhibit, in particular, a diversity of shapes, sizes and substructures that is unparalleled in other land plants (Renzaglia *et al.* 2007). The chloroplast can be spindle- to lens-shaped with a prominent pyrenoid typically surrounded by abundant starch in *Anthoceros*, *Phaeroceros* and *Notothylas*. In *Dendroceros*, it is large and often star-shaped, in *Leiosporoceros*, by contrast, it is rounded with distinct zonation and large, elaborate grana in the central region (Fig. 5.15).

Elucidating the phylogenetic affinities of hornworts based on morphological characters has been hampered by the simple architecture of the gametophyte, lacking much, if any, internal specialization, and by the conflict in phylogenetic signals between the remaining characters. As a result, their classification into genera and taxa of higher rank has undergone many changes, even within the last two decades (cf. Hasegawa 1988, Hyvönen & Piippo 1993 and Duff *et al.* 2007). In previous morphological analyses, *Notothylas* was resolved as sister to the remainder of the hornworts (Hyvönen & Piippo 1993). This is because *Notothylas* shares with algae a series of characters that were interpreted as plesiomorphies, including the presence of a pyrenoid, the short sporophyte of determinate growth, the synchronous spore production, the lack of a columella and the absence of stomata. The latter feature is shared with *Megaceros* and *Dendroceros*, which consequently were basal in the morphological tree.

Analysis of molecular characters resulted in a nearly inverse phylogenetic hypothesis of generic relationships (e.g. Duff *et al.* 2007; Fig. 5.16). Inferences from sequence data suggest that *Notothylas*, *Megaceros* and *Dendroceros*, are, in fact, derived genera and that their unique features represent derived, albeit reduced, characters rather than true plesiomorphies. The molecular phylogeny

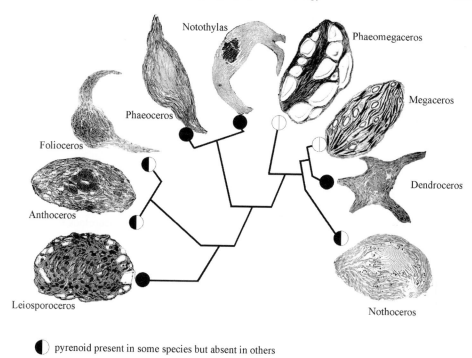

Fig. 5.15. Chloroplast diversity in hornworts, as plotted on a molecular phylogeny of the group (redrawn from Renzaglia *et al.* 2007 with permission of *The Bryologist* and the authors).

suggests for instance that the absence of stomata and pyrenoids result from a loss. Although previously regarded as a useful taxonomic and phylogenetic marker, chloroplast morphology and structure has proved to be highly homoplastic, particularly regarding the recurrent gain and loss of pyrenoids associated with structural modifications of the central chloroplast region (Fig. 5.15).

5.3 Biogeography and ecology

Hornworts are found on all continents except Antarctica. The distribution of most genera spans two or more continents, with four genera (*Anthoceros, Megaceros, Notothylas* and *Phaeoceros*) subcosmopolitan and two genera (*Dendroceros* and *Folioceros*) pantropical in distribution.

All hornworts are ecological pioneers and occur on substrates where they have little competition, mostly on mineral soils but also on rock in streams,

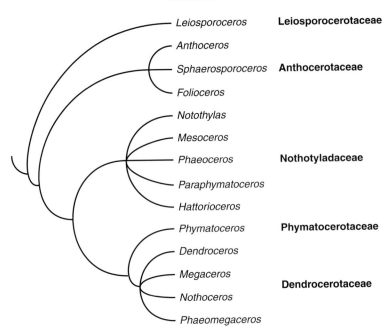

Fig. 5.16. Simplified phylogeny of the Anthocerotophyta as inferred from sequence data (after Renzaglia *et al*. 2009).

trees or even leaves in the tropics. From those primary, natural habitats, hornworts have successfully colonized secondary habitats artificially disturbed by human activities. In Europe, for instance, hornworts are most often, if not almost exclusively, found in stubble fields. This is the case of *Anthoceros neesii*, for example, for which the primary habitat is unknown and has probably been destroyed (Hallingbäck & Hodgetts 2000).

The hornwort life cycle, which is annual in Europe and involves a period of dormancy as spores within the diaspore bank (see Box 7.3) underground (Bisang 1996), is, in fact, well adapted to regularly disturbed habitats. Spores germinate in summer, often right after crop harvest and are thus on bare ground almost free from competition, but still benefiting from the protective cover of remaining crop stubble. The plants die at the onset of winter, which is spent as spores in the soil.

Hornworts exhibit a wide ecological range regarding soil conditions and climatic factors, but are very sensitive to agricultural practices in Europe (Bisang 1998). In particular, ploughing time seems to be the main factor influencing the presence of above-ground hornwort populations. Hornworts are abundant in untreated stubble fields, but become rare or absent from fields immediately ploughed after harvest or heavily fertilized. In the current context of farming policies encouraging ploughing and sowing soon after harvest to enhance soil protection and prevent nitrate leaching, the abundance of hornworts has strongly decreased throughout Europe (Bisang 1992), to the point that some are now endangered. *Anthoceros neesii*, for example, occurred in many central European localities in the past, but is now known at only a few sites in Austria and Germany (Hallingbäck & Hodgetts 2000, Manzke 2004).

6

Biogeography

Biogeography is the science that aims to describe the spatial distributions of biota (a pattern) and understand the means by which these distributions were achieved (a process). Biogeography is a field that existed long before evolutionary biology and indeed helped in founding the evolutionary ideas of Charles Darwin and Alfred Wallace, among others (Humphries & Parenti 1999). Biogeography and evolutionary biology therefore interface with each other, as the discovery of the mechanisms regulating species distributions involves an understanding of species dispersal ability, evolutionary rates and diversification mode, which are among the main foci of the sciences of evolution.

In general, bryophyte species have broad geographic ranges that often span more than one continent (e.g. Figs. 6.1–6.3). Some, termed as 'cosmopolitan', are even widespread across all continents. Bryophyte species thus tend to show wider distributions than vascular plants. In fact, many bryophyte species exhibit the same disjunctions that are well known in flowering plants at the generic level. For example, 43% of the moss species found in North America are also found in Europe, while 70% of the species found in Europe also occur in North America (Frahm & Vitt 1993). By contrast, 48% of the genera, but only 6.5% of the species, are shared between the North American and European vascular flora (Qian 1999). Two competing hypotheses, namely repeated intercontinental dispersal and continental drift, have traditionally been proposed to explain the broad and highly disjunctive distributions typical of bryophyte species.

For Darwinian biologists, biotic distributions were a result of dispersal away from a centre of origin. On an Earth that was thought to be static, they believed the solution to finding why these distributions have occurred was to search for the centre of origin by retracing the steps of genealogical lineages through space and time. They therefore saw each species lineage as

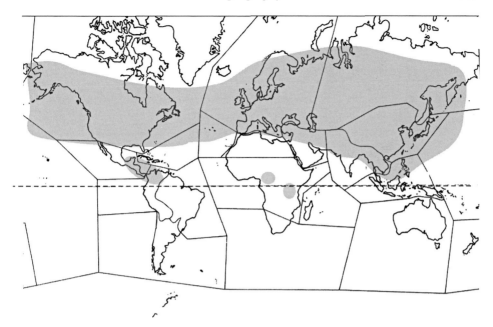

Fig. 6.1. Holarctic distribution of the liverwort *Lepidozia reptans*, with intrusion in high mountain areas in the tropics (redrawn from Tan & Pocs 2000).

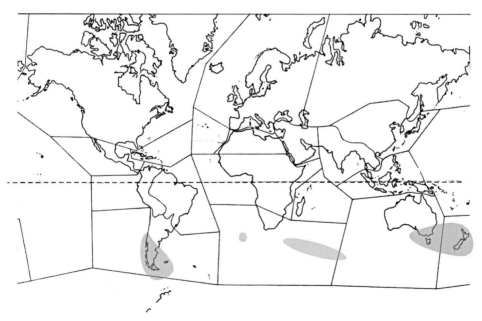

Fig. 6.2. Circum-Subantarctic range of the liverwort genus *Herzogobryum* (redrawn from Tan & Pocs 2000).

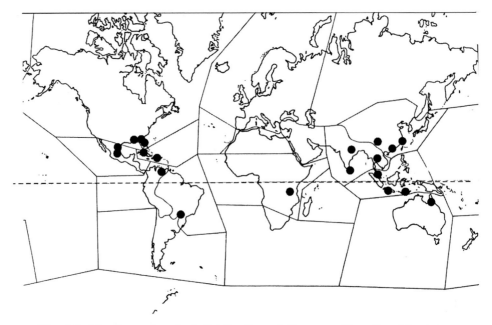

Fig. 6.3. Nearly pantropical distribution of the moss *Erpodium biseriatum* (redrawn from Tan & Pocs 2000).

having its own unique history and extrapolated common biotic distributions to be the result of the same dispersal processes (see Humphries & Parenti 1999 for review).

In the 1960s, the reappraisal of the nature of the Earth's crust by Wegener's plate tectonics theory provided a solid basis for understanding and explaining the floristic similarities among regions of the world that are widely separated today. In contrast to Darwin's theory, dispersal was thought to overwrite, in a stochastic fashion, the initial pattern established by drift. Dispersal, it was argued, is not capable of falsification. As it could explain any pattern, it could conclusively explain no pattern in particular and was regarded as unsuitable for scientific examination. Dispersal was therefore considered an irrelevant noise that could be disregarded in the search for patterns established by drift (McGlone 2005).

It has, however, been increasingly admitted that dispersal patterns may be surprisingly regular and enduring in source, direction and target areas (McDowall 2004). Recent developments based on studies at the molecular level have, in fact, undermined the pre-eminence that continental drift has enjoyed and we are now in the middle of a dispersalist counter-revolution (De Queiroz 2005).

In this chapter, we describe the general patterns of bryophyte distributions. We discuss how these distributions were achieved by contrasting the hypotheses of continental drift and long-distance dispersal. Finally, we discuss the implications that the competing hypotheses of continental drift and long-distance dispersal have for our understanding of morphological evolution and species diversification.

6.1 Bryophyte diversity and distribution patterns

A first goal in biogeography is to document diversity and distribution patterns and determine how those patterns are organized. Takhtajan (1986) was the first plant biogeographer to introduce a comprehensive, detailed system of floral units based on their levels of endemism. Implicitly, the principle underlying Takhtajan's system is that the distinctiveness of each floral unit, as expressed by its level of endemism, is the result of isolation, which has permitted independent evolution within the area in question. Takhtajan recognized six kingdoms in the world floras, namely the Australian, Holarctic, Holantarctic, Neotropic, Paleotropic and Cape, and this system was adopted by Schofield (1992) for the world bryophyte flora (Fig. 6.4). North temperate and boreal areas of North America, Europe, northern Africa, and

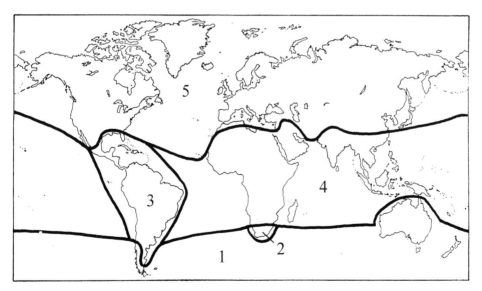

Fig. 6.4. Bryofloristic kingdoms of the world. 1. Holantarctic (including here the Australian kingdom). 2. South African. 3. Neotropic. 4. Paleotropic. 5. Holarctic (redrawn from Schofield 1992).

central and northern Asia compose the Holarctic Kingdom. This kingdom occupies the largest land areas in the world. It is characterized by a series of taxa that are predominantly of Holarctic occurrence, but are not restricted to it (i.e. are not endemic). For example, the liverwort *Lepidozia reptans* (Fig. 6.1) is widespread over the whole Holarctic area, but also occurs in small patches in the high mountains in the tropics. In contrast to the vascular flora, the Holarctic is characterized by very low rates of endemism in bryophytes. For instance, no liverwort genus is endemic to northern Asia, and endemism rates for Europe and North America are only of about 1% (Fig. 6.5).

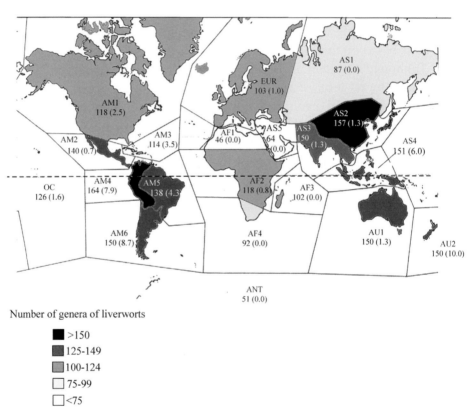

Number of genera of liverworts

- ■ >150
- ▓ 125-149
- ▒ 100-124
- □ 75-99
- □ <75

Fig. 6.5. Number of genera of liverworts and proportion of endemic genera (in parentheses) in each of the 20 world biogeographic regions as defined by van der Wijk *et al.* (1959). Eur = Europe; AF1 = northern Africa; AF2 = continental sub-Saharan Africa; AF3 = Mascarene Islands; AF4 = southern Africa; AS1 = northern Asia; AS2 = central Asia; AS3 = southern Asia; AS4 = southwestern Asia; AS5 = western Asia; AM1 = North America; AM2 = Central America; AM3 = Carribean islands; AM4 = northern South America; AM5 = Brazil; AM6 = southern South America; ANT = Antarctica; OC = Oceania; AU1 = Australia; AU2 = New Zealand.

   The southern hemisphere counterpart of the Holarctic Kingdom is the Holantarctic Kingdom. The latter, merged here with Australia for simplicity, encompasses temperate southern South America, Antarctica, Australia and New Zealand. The genus *Herzogobryum* is, for example, endemic to the kingdom and its species are distributed across the whole area (Fig. 6.2). Rates of endemism in the Holantarctic are among the highest found among the world's bryophytes. For example, 10% and 8.7% of the genera and 52% and 32% of the liverwort species of New Zealand and southern South America, respectively, are endemic. South Africa represents a particular case of an especially rich temperate to sub-tropical Southern Hemisphere bryoflora, wherein more than 300 liverwort (Wigginton 2004) and 500 moss species, including 24% of endemic species, occur (O'Shea 1997).

   The tropical belt spans South America, sub-Saharan Africa, southern Asia and Oceania. Taxa termed 'pantropical', such as the moss *Erpodium biseriatum* (Fig. 6.3), exhibit a distribution range that encompasses all of these areas. The tropical area of the world is subdivided into two components, each representing a floristic kingdom: the Neotropics, composed of Central and tropical South America, and the Paleotropics, comprising the tropical areas of Africa and Asia. Nearly 4000 species of bryophytes (2600 mosses, 1350 liverworts and 30 hornworts), corresponding to about one third of the world's total bryophyte diversity, occur in the Neotropics (Gradstein *et al.* 2001). Within the Neotropics, the northern Andes exhibit the highest species diversity and rates of endemism. About 8% of the liverwort genera and no less than 68% of the moss species are restricted to the area (Gradstein *et al.* 2001). In the Paleotropics, southwestern Asia holds much of the diversity, with about 8% of endemic liverwort genera and no less than 33% of endemic liverwort species in Borneo and New Guinea, respectively (Vanderpoorten & Hallingbäck 2009).

   Nevertheless, in contrast to one of the few truly general patterns in biogeography (Mittelbach *et al.* 2007), the tropical areas of the world are not richer in bryophyte species than temperate areas. As a result, no significant latitudinal gradient of diversity is evident. This is shown, for example, in the examination of worldwide patterns of liverwort genus diversity (Fig. 6.5). Locally, diversity patterns between bryophytes and vascular plants are, therefore, uncoupled. This is, for example, the case in Chile, where species richness patterns for vascular and non-vascular plants are inverted across 25 degrees of latitude (Rozzi *et al.* 2008). The moss flora of tropical lowland forests is, in fact, notably depauperate (Churchill 1998). High levels of bryophyte diversity are found in tropical mountains (Churchill *et al.* 1995, Gradstein 1995), but the numbers presently available for mosses do not support the view that tropical mountains are obviously more species-rich than other regions at higher latitudes (Shaw *et al.* 2005). In the large leafy liverwort families Lophoziaceae and Scapaniaceae, for example, species diversity tends to follow a diversity gradient extending away from the equator (Söderström *et al.* 2007). Although it is true that the current level of floristic knowledge in tropical areas is far below that achieved in temperate areas, the figures would have to change dramatically for a clear trend to emerge (Shaw *et al.* 2005).

6.2 Origin and evolution of bryophyte distribution patterns

6.2.1 Plate tectonics

A next step in the analysis of bryophyte distribution patterns is to understand how these distributions were achieved. In the context of Wegener's continental drift theory, the floristic differences between Southern (Holantarctic) and Northern (Holarctic) Hemisphere flora are assumed to be a reflection of the opening of the Tethys sea some 180 mya, which divided the Pangea supercontinent into two large land masses, Laurasia and Gondwana (Fig. 6.6) (Raven & Axelrod 1974). Fragmentation of Gondwana started *c.* 180 mya and was completed between 80 mya and 53 mya. The very ancient split of Gondwana, with subsequent isolation of southern South America, Australia and New Zealand, may be one of the main reasons why the flora of these areas nowadays exhibit the highest endemic rates worldwide (Fig. 6.5).

By contrast, Laurasia was assembled comparatively late. The opening of the North Atlantic was completed about 60 mya, but land bridges across the Faroes and Greenland still persisted until 40–30 mya, while Eurasia and western North America remained connected via the Bering bridge until 5.5–5.4 mya. These bridges offered an opportunity for migration between Eurasia and North America. For example, amphi-Atlantic ranges (Fig. 6.7), exhibited by 8% of the European and 7% of the North American moss flora and 15% and 12% of the European and North American liverwort flora, respectively, have traditionally been interpreted as a result of the historic connections through the North Atlantic bridge (Schofield 1988). The floras of North America and Europe thus appear strikingly similar, with 68 of 252 moss genera represented by the same species in North America and Europe and interpreted as an old common stock of Laurasian taxa (Frahm & Vitt 1993). As a consequence, rates of endemism in the Holarctic, with, for example, no endemic liverwort genus in northern Asia, one genus (1%) in Europe and three genera (2.5%) in North America, are much lower than in former Gondwanic splinters (Fig. 6.5).

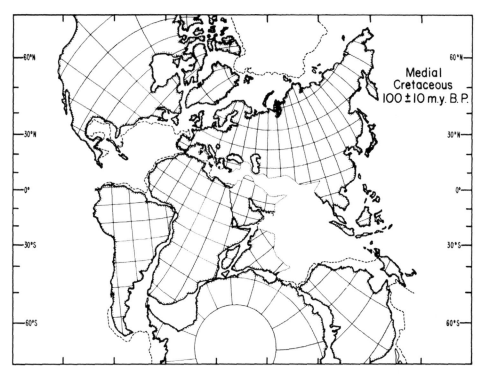

Fig. 6.6. Reconstruction of Laurasia and Gondwana during Medial Cretaceous (100 my BP) (reproduced from Raven & Axelrod 1974 with permission of Missouri Botanical Garden Press).

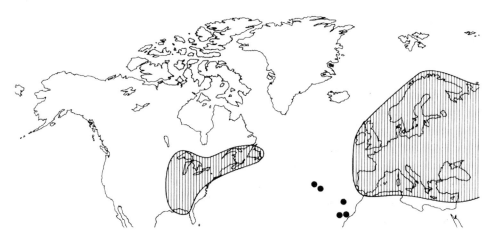

Fig. 6.7. Amphi-Atlantic distribution range of the liverwort *Metzgeria furcata* (reproduced from Schofield 1988 with permission of Blackwell).

6.2.2 Long-distance dispersal

The Earth's history of continental drift provides an adequate explanation for
many of the distribution patterns described above. Yet, dispersal is evident
from contemporary observations of colonization (or recolonization after
major disturbances, such as the oft-cited establishment of biotas on the
Krakatau volcano after the 1883 eruption) of volcanic islands, which are of
relatively recent geological origin and considerable distance from the nearest
sources. The existence of distributions that cannot be merely explained by
continental drift (for example, the disjunct occurrence of the mainly Holarctic
species *Lepidozia reptans* on the highest mountain ranges in the tropics
(Fig. 6.1)), further suggests that at least some of the disjunct distributions
observed today result from long-distance dispersal.

6.2.2.1 Dispersal mode

Bryophytes disperse by means of spores, specialized vegetative propagules
and gametophyte fragments, which are collectively referred to as diaspores.
Dispersal by spores is not an option for a range of species that fail to produce
sporophytes. For instance, 14% of the moss species of the British flora never
undergo sexual reproduction (Longton & Schuster 1983). Dioicous species in
particular often fail to reproduce sexually. In a study on the reproductive
biology of British mosses, Longton (1997) found that 87% of the British
species, wherein sporophytes are unknown, are dioicous, whereas sporophytes
are regarded as occasional to common in 83% of the monoicous species.
Fertilization indeed involves sperm cells swiming to the archegonia, a process
that may be facilitated by micro-arthropods (Box 1.1). Fertilization ranges
are extremely limited. Antherozoids are capable of swimming as much as
1–2 m at a speed of $100–200\,\mu m\,s^{-1}$. The greatest distance between fruiting
plants and the nearest male inflorescence has been recorded at a maximum of
3.8 m in the moss, *Dawsonia superba*. However, fertilization ranges are generally much shorter. Rydgren *et al.* (2006) found, for example, that 85% of the
female shoots with sporophytes were situated within a distance of 5 cm from
the nearest male and the longest distance was 11.6 cm. In monoicous species,
such functional constraint is virtually lacking as the distance between sexes
is minimal. For dioicous species, which represent about 70% of liverworts
and 60% of mosses, the likelihood of fertilization is, by contrast, inversely
proportional to the distance between male and female plants. Fertilization
is, in many instances, complicated by the spatial segregation of the sexes.
Spatial segregation of sexes may result from a sex-specific physiology
(Stark & McLetchie 2006). In the liverwort *Marchantia inflexa*, for example,

males are found under higher light conditions than females (McLetchie & Puterbaugh 2000). In some cases, sexes do not even occur sympatrically (i.e. within the same geographic area), hampering any chance of sexual reproduction. In the hornwort *Megaceros aenigmaticus*, for example, sexes specifically colonize different watersheds (Section 5.1). As an even more extreme example, only male plants of the moss *Syntrichia pagorum* occur in North America, whereas only female plants are known from Europe.

In addition, sexual reproduction incurs a substantially higher 'cost' than asexual reproduction. Sporophyte production is correlated with a higher shoot mortality among females (Pohjamo & Laaka-Lindberg 2003), reduced size of vegetative offshoots, reduced branching (Rydgren & Økland 2003), and reduced production of new reproductive organs (Bisang & Ehrlén 2002). No such trade-off between shoot mortality and gemma production has yet been demonstrated. Sexual reproduction also requires specific habitat and climatic conditions, so that some species freely release spores in some areas, but not in others (Laaka-Lindberg 2000, 2005). For example, Longton (1997) noted that sporophyte production is extremely rare in mosses in polar regions, whereas the same species are abundantly fertile in temperate areas. Consequently, the success of sexual reproduction is highly variable (Rydgren & Økland 2002, Rydgren *et al.* 2006).

Asexual methods of gametophyte reproduction are thus of utmost importance in bryophytes. Some species are only known to reproduce vegetatively. The moss *Fissidens celticus* is, for example, only represented by female plants. Bryophytes display a remarkable variety of mechanisms of vegetative reproduction (see Sections 3.1.3, 4.1.3, 5.1.3). This variety is paralleled by a wide range of cellular separation mechanisms, which are more diverse in bryophytes than any other group of land plants (Duckett & Ligrone 1992).

6.2.2.2 How and how far are bryophytes dispersed?

The deposition curve of diaspores reveals a marked decrease in diaspore density with distance from the source (Fig. 6.8). It is therefore not surprising that a strong tendency for spatial aggregation is found in some species (Fig. 6.9). Using genetic markers, Snäll *et al.* (2004a) showed that individuals of the moss *Neckera pennata* are not randomly distributed across the landscape, as would be expected if the species was not limited by dispersal. Rather, offspring from the same parental source tend to occur most frequently close to the latter. This structure was observed up to a distance of 350 m and further suggests that at least some species have a restricted dispersal range.

Nevertheless, a moss or liverwort sporangium may contain millions of spores and it is likely that some are dispersed at the regional or continental

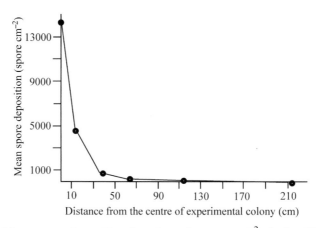

Fig. 6.8. Mean spore deposition (number of spores cm^{-2}) during 30 days as a function of distance from experimental colony in *Atrichum undulatum* (reproduced from Miles & Longton 1992 with permission of *Journal of Bryology*).

scale. Even if a fraction of the spore mass has aborted, as is routinely the case in many mosses (about 20% on average in the moss *Bryum argenteum*), every sporangium has the potential to contribute to regional or transcontinental dispersal. Observations of the colonization of man-made habitats by species far from their nearest natural distribution area confirm that some bryophytes can disperse over fairly long distances. For example, Miller and McDaniel (2004) found that the introduction of concrete walls along a highway in the Adirondack Mountains of New York State increased the bryophyte diversity by 50% above that of native lithic substrata. The composition of the native and mortar floras differed greatly, suggesting that the walls were not colonized by locally abundant ruderal species. Many of the species sampled on the walls were found only in lower elevation forested sites 5 km or more from the roadway and not on concrete. These results suggest that a bryophyte community assembled from sites located at > 5 km at the rate of at least one species per year over the past 65 years. Similarly, Hutsemékers *et al.* (2008) studied slag heap colonization to document ranges and rates of dispersal at the landscape scale and found that 56% of the species that compose the present-day slag heap flora were recruited from source populations located within a range of at least 6–86 km within a period of less than fifty years.

    If these observations are now projected on to the scale of centuries rather than decades, long-distance dispersal events may play an important role in the establishment of global patterns of species diversity, provided that spores remain viable during transportation. The germination capacity of non-aborted spores at the time of spore liberation is generally high. Söderström and Herben (1997) reported figures above 90% for a wide range of mosses.

Fig. 6.9. Occurrence of the obligate epiphytic moss *Neckera pennata* on a background of suitable past and present broadleaf stands in a Swedish boreo-nemoral landscape (reproduced from Snäll *et al.* 2004b with permission of Blackwell).

The longevity of spores varies, however, among species. Spores of some leafy liverworts survive only a few hours. Only 20% of spores of *Buxbaumia viridis* germinate after 12 days of dry storage, compared to 60% in *N. pennata*. Virtually all spores of both species died after 48 days of storage (Fig. 6.10) (Wiklund & Rydin 2004a). In *Funaria hygrometrica*, viability of spores stored at room temperature in a capped jar fell from 100% to 5% after 11 years, while spores of thalloid liverworts of xeric habitats remain viable for several years. These tolerances were, however, measured in experimental conditions and it is very likely that viability would be much lower in field conditions due to, for example, predation or decay.

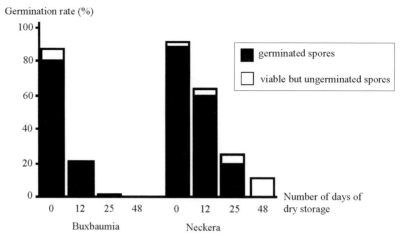

Fig. 6.10. Germination of spores of *Buxbaumia viridis* and *Neckera pennata* after 12, 25 and 48 days of dry storage on dry filter paper. Black parts of bars show germinated spores, whilst white parts show ungerminated, but still viable spores (reproduced from Wiklund & Rydin 2004a with permission of Blackwell).

Such differences in spore viability may explain the difference in geographical ranges among species. In fact, a marked correlation between species range and survival of spores under the conditions of desiccation and frost that prevail in high altitude air currents was found among 139 species displaying contrasting distribution ranges (van Zanten 1978). The spores of New Zealand endemics survived desiccation for about half a year, but were killed by desiccation followed by dry freezing. Both the period over which spores survived when dry and the ability to germinate following desiccation and frost were shown to increase progressively along the following sequence of distribution patterns: New Zealand, New Zealand+Australia/Tasmania, Australasia+South America, Australasia+Northern Hemisphere (Fig. 6.11). These experiments show that the spores of species with broad ranges can tolerate the extreme conditions at high altitudes and offer one explanation for the difference in the distribution ranges observed: endemic species remain restricted to a specific area because their spores do not survive trans-oceanic dispersal in high-altitude air currents. By contrast, species equipped with drought- and cold-tolerant spores have the potential to exhibit intercontinental disjunctions.

Most bryophyte species are indeed wind-dispersed. This has long been acknowledged based on empirical evidence, but Muñoz *et al.* (2004) were the first to demonstrate that bryofloristic patterns exhibit a striking parallelism with wind connectivity rather than geographic proximity (Fig. 6.12).

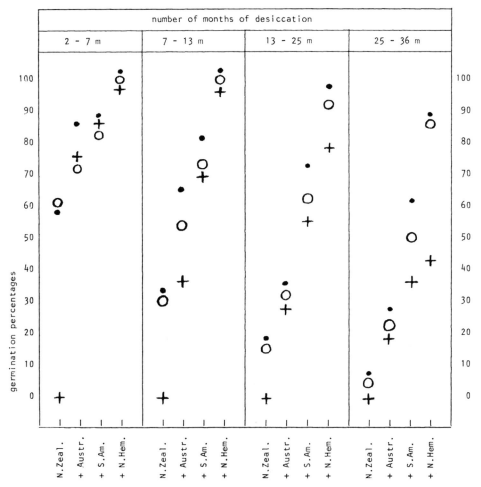

Fig. 6.11. Percentages of germination of spores exposed to desiccation at ambient room relative humidity during 2–36 months (•) followed by subsequent wet freezing during four days at –30°C (○) or followed by subsequent dry freezing during four days at –30°C (+) in 139 moss species with contrasting geographic ranges, including endemics to New Zealand; New Zealand+Australia/Tasmania; New Zealand+Australia/Tasmania+ southern America/Subantarctic islands; New Zealand and Northern Hemisphere (reproduced from van Zanten 1978 with permission of the Hattori Botanical Laboratory).

Zoochory (i.e. transportation by animals) has often been mostly considered a curiosity restricted to the moss family, Splachnaceae. In the latter, dispersal is mediated by flies, which are attracted to the strongly modified capsules by a series of chemical compounds similar to those produced by insects themselves (Box 4.2). The significance of zoochory for bryophyte dispersal has,

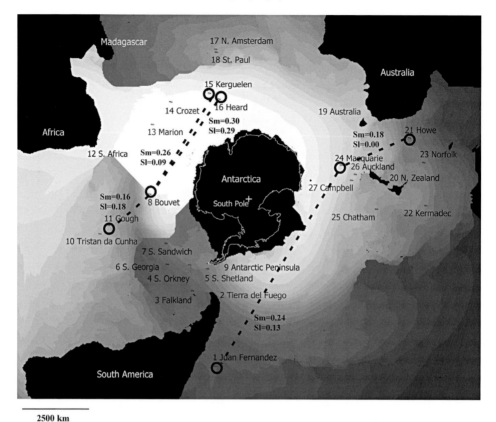

2500 km

Fig. 6.12. Intensity of wind connectivity (increasing from dark to pale tones) in the Subantarctic area for the period from 1 to 10 February 2002. Bouvet Island exhibits a higher floristic similarity for mosses (Sm) and liverworts (Sl) with the geographically distant Kerguelen (4420 km) and Heard (4430 km), with which wind connectivity is higher, than with the closer Gough island (1860 km), with which connectivity by wind is low. A similar case is illustrated by Macquarie, Howe and Juan Fernandez Islands (reproduced from Muñoz *et al.* 2004 with permission of *Science*).

however, been increasingly acknowledged. Despite the lack of specific means of ensuring attachment, like seed ornamentation in angiosperms, the adhesive transportation of unspecialized bryophyte fragments is likely to be enhanced by various morphological characters, including a small size, numerous branches and erect or squarrose leaves. Epizoochory (transportation on animal bodies) seems especially important in habitats such as woodland, where animal dispersal agents are common and wind velocity is low. In such habitats, the significance of epizoochorous transport of bryophytes seems to be actually comparable to that of phanerogams. In 30 random trials on roe deer, for example, Heinken *et al.* (2001) collected 51 bryophyte fragments against 1969 angiosperm diaspores, which was roughly proportional to the 1:30 bryophyte-herb ratio prevailing on the ground. Similarly, forest birds foraging on the ground are known to disperse bryophyte fragments over short distances (< 2 m) (Davison 1976). The correlation between the distribution of some species, such as *Timmia megapolitana*, and bird migratory routes, further suggests that birds may also play an important role in long-distance dispersal (Porley & Ellis 2002). In fact, recurrent observations of fragile shoot apices attached to bird feet were made, for example, on skuas and penguins in Antarctica (Lewis Smith 1999). Accidental ingestion may also occur, so that viable fragments can be dispersed over long distances via passage through the digestive tract of highly mobile vertebrates such as flying foxes and birds (endozoochory) (Proctor 1961, Parsons *et al.* 2007).

6.2.2.3 The influence of humans in bryophyte distribution patterns

The introduction of species outside their primary distribution range by humans represents an artificial case of long-distance dispersal. A plant is alien (neophyte) if its presence in a given area is due to intentional or accidental introduction by humans. If it reproduces consistently and sustains populations over many life cycles without direct human intervention, it is termed 'naturalized' (Richardson *et al.* 2000).

Bryophytes are never deliberately introduced, except in moss gardens. Fossils, whose presence in the geological layers would immediately exclude the hypothesis of introduction, are exceedingly scarce. Hence, only indirect evidence can be used to detect if a species is native or introduced (Söderström 1992).

i) Expansion and/or anomalous geographical distribution. The moss *Orthodontium lineare* has, for example, a primarily circum-sub-Antarctic distribution but was introduced in Europe (Fig. 6.13) (Söderström 1992). Range expansion is, however, not a decisive criterion in itself because native species can also exhibit similar trends. For example, the moss *Pogonatum dentatum* used to be a mountain plant in Scandinavia but started to spread southwards during the second half of the twentieth century, probably using forest roads as dispersal corridors over long distances (Hassel *et al.* 2005).

ii) Association with some means of introduction, for example, botanical gardens or ports. The Southern Hemisphere liverwort *Lophocolea semiteres* is now present and sometimes locally abundant in the UK, The Netherlands and Belgium. It was initially found in gardens on the Isles of Scilly, off south-west England, where it is thought to have been introduced in the early twentieth century with garden plants from Australia (Stieperaere 1994).

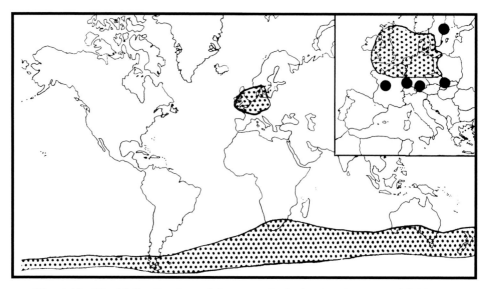

Fig. 6.13. World distribution of the moss *Orthodontium lineare*, exhibiting a primary Subantarctic range with subsequent introduction and spread in Europe (redrawn from Söderström 1992).

iii) Association with open, disturbed or temporary sites. For example, *Pseudoscleropodium purum*, a moss native to central and western Europe, is well established in New York State, where its restricted occurrence to lawns of cemeteries suggests that the spread of the species has been achieved through horticultural practices such as mowing, raking and planting (Miller & Trigoboff 2001).

All these factors provide, however, rather inconclusive indications of introduction and it might be difficult to distinguish between natural long-distance dispersal and accidental introduction. For example, *Atrichum crispum*, a common species in eastern North America, was first detected in Wales in 1848, where it grows on sand or gravel near water, a substratum characterized by a natural disturbance regime. All British populations are male, which suggests a single successful introduction of this dioicous moss. Whether the species was naturally introduced from a wind-borne spore or human transportation is, however, unknown (Söderström 1992).

Introduced species that produce offspring in large numbers and at substantial distances from parental plants have the potential to spread over a considerable area and thus to invade natural, semi-natural or human-made ecosystems. Three species among the 22 introduced in the European moss flora have spread and can be termed 'invasive' (Söderström 1992). One of these species, *Campylopus introflexus*, which is primarily distributed in the

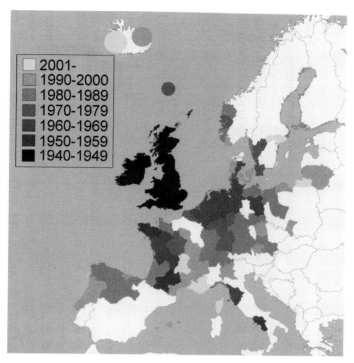

Fig. 6.14. Timing of invasion of Europe by the Southern Hemisphere moss *Campylopus introflexus*, as shown by the time of first report in each country (reproduced from Hassel & Söderström 2005 with permission of the Hattori Botanical Laboratory).

Southern Hemisphere, has been found in Europe, according to the oldest collection of the species, made in the UK, since the 1940s. The sequential map of the spread (Fig. 6.14) shows a rapid expansion from northwestern Europe eastwards and southwards (Hassel & Söderström 2005). The species now occurs throughout Western and Central Europe and is, at least locally, very abundant and can become a threat to the local flora (Section 2.2).

6.3 Evolutionary significance of bryophyte distribution patterns

Accepting either continental drift or long-distance dispersal as explanatory factors for the disjunctions observed in the range of many bryophyte species has major evolutionary consequences. Many bryophyte species exhibit the same disjunctions that are well-known in angiosperms at the generic or family level and this pattern has sometimes been interpreted as evidence for a common historical explanation, i.e. ancient vicariance associated with

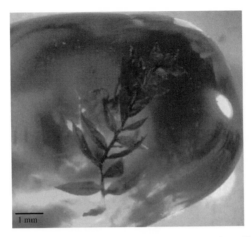

Fig. 6.15. Baltic amber fossil of the moss *Trachycystis flagellaris* (37–57 mya) (reproduced from Frahm 2004a with permission of Elsevier). See plate section for colour version.

continental movements and global climate change during the Tertiary Period (see Shaw 2001 for review).

In some instances, fossil evidence unambiguously testifies to the widespread occurrence of taxa, nowadays extinct, in the Tertiary flora. The best source of fossil information for bryophytes comes from tissues preserved in Baltic, Saxon and Dominican amber (Frahm 2000, 2004a). Amber is a fossil resin originally produced by a conifer, *Pinus succinifera*, growing in oak-pine forests in Fennoscandia during the Eocene Period (57–37 mya) in the case of Baltic and Saxon ambers and by an angiosperm, *Hymenaea* sp., during a period estimated to range from the Eocene to the Oligocene Period in the case of Dominican amber (Frahm 2004a, Frahm & Newton 2005). In contrast to fossils in other deposits, tissues included in amber are usually perfectly conserved and are therefore important sources for the reconstruction of fossil floras (Fig. 6.15). For example, the liverwort genus *Nipponolejeunea* is currently restricted to Japan. However, the presence of amber fossils of *N. europaea*, which are morphologically almost identical and might even be conspecific with the Asian species, unambiguously attest that the distribution of the genus was much broader in the early Tertiary Period (Schuster 1983) (Fig. 6.16). *Nipponolejeunea* is therefore termed a paleoendemic genus because its endemism results from the contraction of a formerly larger distribution range. Other examples of Baltic and Saxon amber fossils of mosses that are nowadays restricted to eastern Asia support the hypothesis that the distribution of these taxa once spanned Eurasia (Frahm 2000, 2004a).

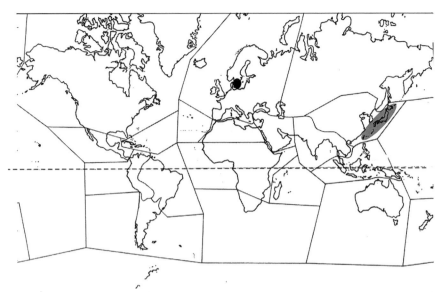

● Extant range of *Nipponolejeunea subalpina*
● Oligocene fossils of *N. europaea*

Fig. 6.16. Extant range of *Nipponolejeunea subalpina* and Oligocene fossils of *N. europaea* (redrawn from Schuster 1983).

The corollary of such an interpretation is that bryophytes with disjunct distributions are extremely old. For example, an amphi-Atlantic distribution suggests, under the assumption of vicariance, that the species has undergone little or no morphological evolution in the 60 myr time span that has elapsed since the North Atlantic opened (Schuster 1983). It is such observations that have led to the traditional view of bryophytes as unmoving, unchanging relics of the past (Crum 1972). In fact, whilst a few bryophyte fossils exhibit a morphology that is unknown among extant taxa (Ignatov 1992), many of the moss and liverwort fossils from the Eocene and Pliocene Periods have been attributed to extant species or are morphologically similar to them (Gradstein 1993, Frahm 2004a, Frahm & Newton 2005). Fossil evidence from earlier periods is, by contrast, exceedingly scarce. The fossil record for hornworts is particularly depauperate, in part because of the fragile nature of the thallus. The oldest hornwort fossil dates back to the Cretaceous, about 300 mya. The fossil strongly resembles extant taxa in its spore ornamentation.

In mosses, *Campylopodium allonense*, which dates back to the late Cretaceous Period (Konopka *et al.* 1998), is one of the rare pre-Tertiary fossils, for which the presence of sporophytes allows a satisfactory taxonomic placement. The fossil exhibits several characters that indicate a relationship to the Dicranaceae (Fig. 6.17). The close morphological similarity between *C. allonense* and extant Dicranaceae suggests, together with the few other unambiguous fossil records (e.g. *Eopolytrichum antiquum*, Konopka *et al.* 1997) and a few Cretaceous hornwort fossils, that some groups of bryophytes have persisted with very few morphological changes for at least 80 myr.

Fossil evidence is, however, inconclusive in many instances. The fossil record for bryophytes is extremely incomplete, whether through low probability of fossilization due to habitat preferences or poor preservation due to absence of cutinized or woody tissue. Even in amber, the percentage of bryophytes is very low. Ninety-nine per cent of all fossils from Baltic amber consist of arthropods and, in the remaining per cent of plant fossils, bryophytes are extremely rare. To date, only about 100 species are known from the fossil European bryophyte flora (Frahm & Newton 2005). The question of whether some fossils truly correspond to extant taxa, and even of whether they actually belong to bryophytes, also remains open because gametangia and sporophytes are most rarely available (Schofield 1988).

Because of these limitations, the origin of the transcontinent disjunct distributions typical of bryophytes remains an area of controversy. Statistical techniques combining phylogenetic information and distributions of extant taxa have therefore been developed to disentangle ancient vicariance from dispersal by contrasting the observed distributions with expectations if continental drift were actually the main factor (Box 6.1). Most recently, molecular dating (see Section 1.3.2) has provided new insights to revisit the traditional conflict between ancient vicariance and long-distance dispersal and its evolutionary consequences. The results of molecular techniques support, in some instances, the hypothesis of morphological stasis in transcontinental bryophyte species. Even though it seems that mosses display the lowest rates of molecular evolution among land plants (Stenoien 2008), the idea that such a stasis in morphological traits is paralleled by a complete absence of evolution at the sequence level is, by contrast, consistently refuted. For example, substantial divergence in DNA sequences was found between southern South American and Australasian populations of the trans-Antarctic moss *Pyrrhobryum mnioides* (McDaniel & Shaw 2003). The age of the trans-Antarctic disjunction was estimated at 80 mya, which is consistent with the hypothesis of a Gondwanan vicariance. The complete absence of morphological differences among populations that have diverged for the last 80 myr is a spectacular example of 'cryptic speciation', i.e. the accumulation of genetic differences that is not paralleled by morphological transformations. In the most extreme cases, genetic differentiation extends beyond the morphospecies

Fig. 6.17. Comparison of fossil (1) *Campylopodium allonense* and extant (2) *C. medium* and (3) *Dicranella cerviculata* Dicranaceae. (a) Capsule, lateral view (note overall similarity in capsule shape). (b) Base of the capsule, showing attachment of twisted seta and presence of struma in both fossil and extant plants. (c) Lateral view of peristome (note shared compound annulus and divided tips). (d) Abaxial surface of peristome teeth (note shared vertical striations and divided tips). (e) Adaxial surface of peristome teeth (note shared asymmetrical trabeculae and vermiform sculpture). (f) Compound annulus (reproduced from Konopka *et al.* 1998 with permission of *American Journal of Botany*).

Box 6.1

**Cladistic biogeography, or how to retrace the origin
and evolution of species distributions**

Historical biogeography has traditionally been seen as a series of different propositions for the causes of present day distribution patterns. For Buffon's patterns of allopatric distribution, de Candolle suggested that earth history was all important. For Humboldt, former land bridges explain disjunctions. For Wallace and many early twentieth century workers, dispersal by Darwin's 'occasional means of transport' played a significant role. The problem with all the solutions is that they lack a testable methodology. This is the main motivation behind the development of cladistic biogeography. Cladistic biogeography is about understanding relationships of areas through discovery of biotic patterns (Humphries & Parenti 1999). Systematic patterns become comparable and understandable for biogeography when expressed as area cladograms. Area cladograms express relationships of areas as determined from phylogenies by substituting the name of a taxon for the area in which it occurs. The methodology of cladistic biogeography is still developing and we will take the example of a technique called parsimony-based tree fitting to illustrate it. The method attempts to fit an organismal phylogeny with terminal taxa replaced by geographical ranges (the taxon area cladogram, TAC) (Box 6.1 Fig. 1a) and a general area cladogram (GAC) describing the historical relationships between the geographical areas (Box 6.1 Fig. 1b). Usually, the latter is derived from geological information, such as the chronological sequence of continental split and drift. The fitting procedure produces a reconstruction that specifies a set of ancestral distributions and biogeographical events that, given the GAC, provide an explanation of the observed distribution of the lineage (see Sanmartin *et al.* 2007 for review). Four types of events, including vicariance, speciation within an area, extinction and dispersal, are considered. Each event is associated with a cost and the overall cost of the reconstruction is simply the sum of costs for the inferred events. In parsimony-based tree fitting, the cost of all possible competing scenarios is computed in search of the most parsimonious solution, i.e. the one with the lowest cost. For example, consider the TAC in Box 6.1 Fig. 1a, with five species (1–5) occurring in four areas (A–D) and the GAC depicted in Box 6.1 Fig. 1b. In this case, three vicariance events, one duplication and one extinction are required to explain the observed distribution pattern.

Such a technique was, for example, used to determine the origin of the North Atlantic range disjunction observed in the moss genus *Homalothecium*, which exhibits a typical Mediterranean disjunction as depicted in Fig. 6.19. Several competing hypotheses, including the split from a common broad-leaved evergreen sclerophyllous vegetation following the opening of the North Atlantic

Box 6.1 (cont.)

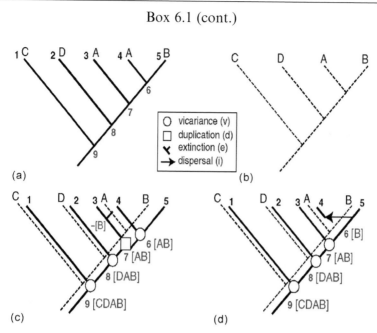

Box 6.1 Fig. 1. Parsimony-based tree fitting. (a) An organismal phylogeny of five taxa (1–5) distributed in four geographic regions (A–D).
(b) A general area cladogram describing relationships between the areas.
For example, C, D, A, B are fragments of a former super-continent
ABCD that successively split and drifted apart. Parsimony-based tree
fitting is a method that aims to reconcile the species phylogeny and
the general area cladogram by minimizing the number of events of
vicariance (v), extinction (e), dispersal (i) and sympatric speciation (d).
(c, d) Two potential solutions to fitting the taxon area cladogram to the
general area cladogram. With the cost assignment procedure described
in Sanmartin *et al.* (2007), the cost of scenario (c) is $1d + 3v + 1e = 1.04$.
(d) The alternative scenario (d) has a cost of $3v + 1i = 2.03$, and is thus
less parsimonious than the scenario depicted in (c) (reproduced from
Sanmartin *et al.* 2007 with permission of Blackwell).

Bridge, the dispersal via the North Atlantic or the Bering Land Bridges coupled
with subsequent regional extinctions, or more recent trans-oceanic long-distance
dispersal, have commonly been proposed to account for the modern distribution
of these Mediterranean floras. Each *Homalothecium* species was assigned one or
several of four distribution areas, including eastern and Western North America,
as delimited by the mid-continental seaway that separated the continent
longitudinally in the late Cretaceous Period, western Palearctic (i.e. western
Europe, western Asia and northern Africa), and Macaronesia. Three different
scenarios were tested for explaining the taxon area cladogram obtained
(Box 6.1 Fig. 2).

Box 6.1 (cont.)

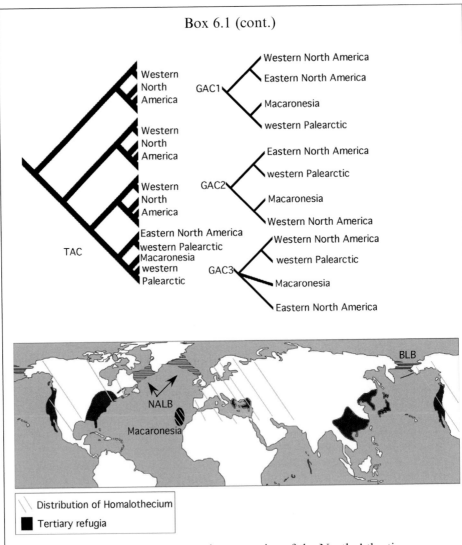

Box 6.1 Fig. 2. Testing competing scenarios of the North Atlantic range disjunction in the moss genus *Homalothecium*. The simplified taxon area cladogram (TAC) is compared to three competing general area clado-grams (GAC) to explain the current distribution of the genus: (1) the Cretaceous mid-continental seaway serves to explain the eastern–western North American disjunction, but the main barrier is the North Atlantic Ocean (Nearctic hypothesis); (2) the North Atlantic Land Bridge (NALB) hypothesis proposes that western Europe and eastern North America were still connected until about 15 mya; the floras of Macaronesia and western North America share similarities because these areas served as refugia during the Tertiary; (3) the Bering Land Bridge (BLB) hypothesis proposes that Western North America and western Palearctic were still connected until about 5 mya. In this scenario, the extinction of the genus must be hypothetized in the eastern Palearctic area.

Box 6.1 (cont.)

The alternative scenarios had, unfortunately, the same cost. It must be borne in mind, furthermore, that even distributions that seem consistent with the continental drift model may actually have been generated by dispersal. Therefore, many molecular biogeographers now utilize a temporal congruence approach, wherein a vicariance event is identified when the divergence time for two lineages coincides with the age of the barrier that separates them (Upchurch 2008). Molecular dating is the process by which an age is placed on a divergence event, represented by a node on a phylogeny, indicating when two populations, species or clades of species, diverged from one another (Section 1.3.2). In the moss genus *Homalothecium* for example, species from North America and Europe are reciprocally monophyletic (Box 6.1 Fig. 2). This observation seems at first sight consistent with the hypothesis that the two lineages diverged following the opening of the North Atlantic Ocean (60 mya) and the sundering of the North Atlantic Bridge, which linked the two continents until about 20 mya. Using molecular dating techniques, Huttunen *et al.* (2008) estimated the divergence of the American and European *Homalothecium* clade at 4.8 mya. This time-scale is not compatible with an ancient vicariance hypothesis and it must be assumed that the disjunct distribution and subsequent species diversification observed in the genus were achieved by recent long-distance dispersal.

---

level, so that morphologically identical taxa do not share a common ancestor. In the moss genus *Mielichhoferia*, for example, some specimens of *M. mielich-hoferiana* are genetically more closely related to *M. elongata* than to other *M. mielichhoferiana* specimens (Fig. 6.18) (Shaw 2000).

In other instances, molecular dating suggests that distributions consistent with the continental drift model may reflect a complex mix of relictualism overlaid by more recent evolution and dispersal. In fact, long-distance dispersal is a much more important factor for explaining the disjunct distributions of bryophytes than previously thought. For example, the hypothesis according to which the Mediterranean disjuncts between the Pacific coast of North America and southern Europe (Fig. 6.19) reflects fragmentation of an ancestral, widespread flora that existed across western North America and Eurasia under sub-humid conditions until the end of the Oligocene Period, about 25–20 mya, has been challenged. The extremely low divergence of DNA sequences found between North American and European populations is not compatible with a divergence dating back to the sundering of the North Atlantic Bridge and rather indicates that recent long-distance dispersal is a

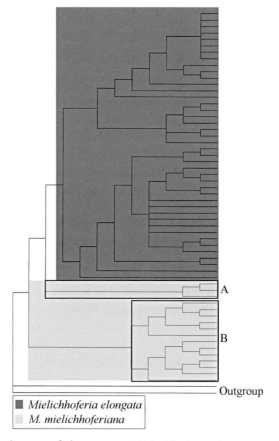

Fig. 6.18. Phylogeny of the mosses *Mielichhoferia elongata* and *M. mielich-hoferiana* based on the analysis of nuclear DNA sequences, showing that the populations of *M. mielichhoferiana* belong to two lineages A and B. Lineage A shares an ancestor with *M. elongata* and is thus more closely related to that species than to populations of *M. mielichhoferiana* of clade B (redrawn from Shaw 2000).

more likely explanation for intercontinental disjunctions in these species (Shaw *et al.* 2003, Huttunen *et al.* 2008).

Based on the analysis of genetic differentiation among bryophyte populations, it is evident that most species are capable of efficient dispersal (see Korpelainen *et al.* 2005, for a review). For example, populations sampled from four vegetation zones across Scandinavia of the widespread moss *Hylocomium splendens* exhibited extremely low levels of relative differentiation among them (Cronberg *et al.* 1997). Similarly, almost no genetic differentiation was found among populations of most *Polytrichum* species at the continental scale, indicating a considerable degree of gene flow by wind-dispersed spores over substantial distances (van der Velde & Bijlsma 2000).

Fig. 6.19. Distribution range of the moss *Antitrichia californica*, an example of disjunction between Mediterranean areas of the Pacific Northwest and southwestern Europe (reproduced from Schofield 1988 with permission of Blackwell).

One consequence of the intensity of gene flow among bryophyte populations is that it may prevent genetic isolation and speciation. In the moss *Pleurochaete squarrosa*, continental and island populations share the same pool of genetic variation, leading Grundmann *et al.* (2007) to propose that large Mediterranean islands function as mainland for bryophytes. That is, the sea does not constitute a barrier to gene flow, hampering any possibility of endemic speciation on islands. The strong gene flow among bryophyte populations might be one reason why rates of endemism in the bryophyte flora are, comparatively, substantially lower than in the angiosperm flora. For example, the west Mediterranean islands Corsica and Sardinia lack any endemic bryophytes, whereas they harbour 10% of endemic angiosperm species (Sotiaux *et al.* 2007). In the mid-Atlantic archipelagos of Macaronesia (the Azores, Canaries and Madeira), which are noteworthy for their 40% endemism rate among angiosperms, bryophyte endemism falls below 10% and is one of the lowest among all other biota (Vanderpoorten *et al.* 2007).

Some cases of fast speciation within a short time-frame, known as 'radiations', have been, however, documented in mosses. For example, molecular dating techniques showed that the moss genus *Mittyrhidium* originated approximately 8 myr ago on Peninsular Malaysia. From there, the group diversified in both easterly and westerly directions to islands in the Indian and Pacific Oceans at a rapid diversification rate of 0.56 ± 0.004 new lineages per million years, which is comparable with other key examples of rapid endemic speciation like, for example, the silversword alliance in Hawaii (Wall 2005). Such cases of fast radiation, which sharply contrast with the complete morphological stasis documented in other mosses, raise the question of why some genera diversify so rapidly, whereas in other taxa no transformation happened over tens of millions of years.

Endemic radiations in other organisms such as pteridophytes have consistently been associated with allopolyploidization. Indeed, allopolyploids are thought to perform better than their progenitors because increased genetic diversity is likely to result in the formation of new, adaptive gene combinations. On the basis of chromosome numbers, it is estimated that 79% of mosses, 11% of liverworts and 2% of hornworts are polyploid, and an increasing body of literature suggests that allopolyploidy is a common speciation process in mosses (Shaw 2009). However, as opposed to pteridophytes, the link between high ploidy levels and successful radiation has not yet been demonstrated in bryophytes.

One explanation is that certain characteristics of the organism, such as reproductive life history strategies, represent key innovations that could be responsible for the rise in diversity. *Mitthyridium*, for example, is diverse in reproductive behaviour, ranging from fully sexual to simply vegetative, lacking any sign of reproduction beyond regular gametophytic growth. *Mitthyridium*, like other mosses (Stark *et al.* 2000), is able to shift reproductive strategies under different environmental circumstances. Such a shift, if fixed in a population, could be a powerful creative force in evolution.

Plate 1.1a

Plate 1.1b

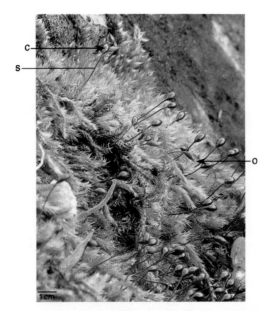

Plate 1.2

Plate 1.3

Plate 2.1

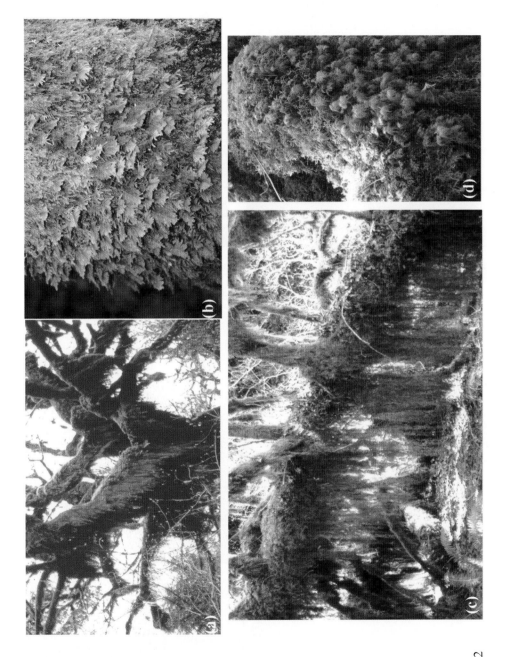

Plate 2.2

Plate 2.7a

Plate 2.7b

Plate 2.7c

Plate 2.9

Plate 2.6

Plate 3.20

6 mm

(a)

1 cm

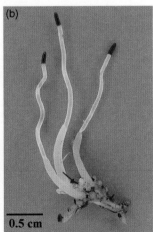

(b)

0.5 cm

Plate 3.21

Plate 3.18

Plate 3.25

Plate Box 4.2

(a)   (b)   (c)
20 µm   250 µm   1 mm

Plate 4.21

Plate 5.8

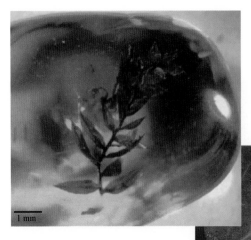

Plate 6.15

Plate 7.6

Plate 8.4

Plate 8.6   1 cm

Plate 8.7

Plate 8.12

Plate 9.7

Plate 9.9

Plate 10.6

Plate 10.3a

Plate 10.3b

Plate 10.3c

Plate 10.10

Plate 10.11

7

Ecology

In the previous chapter, we focused on the mechanisms that explain how bryophyte species are able to 'travel' across oceans and continents. The success in colonizing a new area, or the ability to persist within the same area in the long-term, depend, of course, on the ability of the species to cope with local environmental conditions. Bryophyte distributions are influenced by a variety of factors operating over a range of temporal and spatial scales (Rydin 2009) (Fig. 7.1), which are the focus of the present chapter.

7.1 Global ecology

A central thesis to plant ecology is that climate exerts the dominant control on the distribution of plants at the continental scale. In bryophytes, changes in communities depending on major climatic shifts during the Quaternary Era are evident from the succession of macro-remains preserved in peat. Each climatic phase had its characteristic range of habitats and these appear to have been colonized by bryophytes at a remarkable speed as soon as they developed and became available (Jonsgard & Birks 1995). Two macrofossil taxa are of particular significance in the interpretation of macro-climatic changes: *Sphagnum* spp. and *Racomitrium lanuginosum*.

The local distribution of *R. lanuginosum* is controlled by its preference for cold, wet climates (Tallis 1995). By contrast, the occurrence of *Sphagnum* largely depends on water availability. In the Holarctic, for example, *Sphagnum* bogs are strictly restricted to areas experiencing an annual moisture balance > 0 (Gignac *et al.* 2000). The annual moisture balance measures the difference between mean annual total precipitation and evapotranspiration, which is directly related to the mean annual temperature. The annual moisture balance is thus > 0 when precipitation is more important than evapotranspiration. The distribution of *Sphagnum* bogs is particularly sensitive to

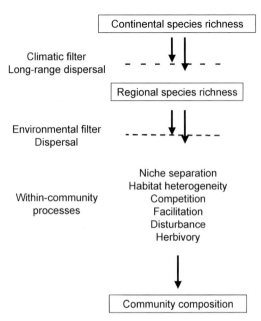

Fig. 7.1. Nested spatial structure of the processes shaping bryophyte species diversity and composition (reproduced from Rydin 2009 with permission of Cambridge University Press).

the moisture balance because their moisture is derived entirely from precipitation. *Sphagnum* bogs are therefore also called ombrotrophic bogs. Because bryophytes lack roots, they cannot move water upwards when the water table is some distance below the surface and are hence highly sensitive to drought periods during the growing season. As a result, the distribution of *Sphagnum* bogs is controlled by the cool temperatures and high precipitation levels that are characteristic of boreal, subarctic, cool temperate and cool oceanic areas (Fig. 7.2).

From the analysis of the proportion of *Sphagnum* and *Racomitrium* remains along a radiocarbon-dated stratigraphic peat profile, Ellis and Tallis (2000) inferred a succession of periods of wetness and coldness separated by warmer and drier periods (Fig. 7.3). The timing of wetshifts in the stratigraphic record was compared to known patterns of climate change derived from instrumental temperature and rainfall data as well as documentary records for the last 900 years (Fig. 7.3). Wetshifts inferred from an increase in *Racomitrium* in the fossil record clearly match documented periods of increased wetness and coldness, which followed the early warm Medieval Period (i) and recurred during the Little Ice Age (ii).

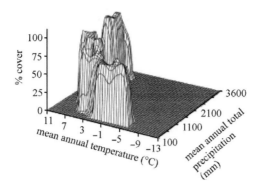

Fig. 7.2. The abundance of *Sphagnum* species in peatlands of western Canada in relation to mean annual temperature and mean annual total precipitation (reproduced from Gignac *et al.* 2000 with permission of Blackwell).

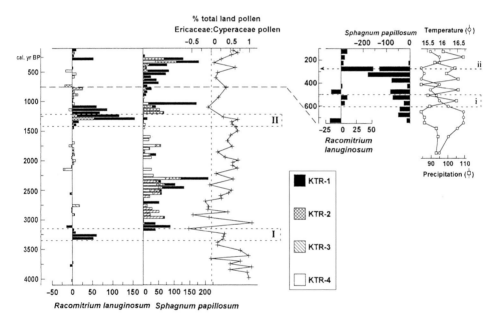

Fig. 7.3. Radiocarbon-dated stratigraphic change (percentage frequency of occurrence) in macro-remains of *Racomitrium* and *Sphagnum* at four sampling sites (KTR1–4) in a British peat bog (left) and summary diagram comparing the timing of inferred wetshifts from the peat stratigraphy with July and August rainfall values and yearly mean temperatures with selected paleoclimatic events (i and ii) indicating periods when increased climatic wetness/coldness has impacted on the local vegetation history (right). The data validate evidence for two earlier major wetshifts (I and II), between which *Sphagnum* outcompeted *Racomitrium* (reproduced from Ellis & Tallis 2003 with permission of *Journal of Bryology*).

The signature of past climatic fluctuations can also be found in the genetic structure of extant populations. For instance, the Quaternary Era was marked by a succession of dramatic cold periods, during which the population of a species either succeeded in migrating southwards to escape the cold front, or disappeared. A good example is found in *Sphagnum fimbriatum* and *S. squarrosum* (Szövényi *et al.* 2006, 2007). The populations of both species dramatically decreased during the glaciations. As a result, extant populations originate from the recolonization of a subset of populations restricted to refuge areas during the glaciations. Their genetic diversity is, hence, fairly low, but examination of their genetic structure reveals that the two species experienced contrasting histories of recolonization. The populations of *S. fimbriatum* are genetically highly structured, suggesting a complete lack of genetic interchanges among populations during the glaciations and recolonization from a small number of refugia along the Atlantic coast. By contrast, the genetic variation of *S. squarrosum* exhibits almost no spatial structure across Europe, suggesting that the species was less affected by the glaciations than *S. fimbriatum* and survived in more numerous refugia, from which populations spread after climate change.

The significance of macroclimatic variation in shaping bryophyte distributions has major consequences for interpreting or predicting past, present and future distributions. On the one hand, a consequence of the response of bryophyte communities to climatic variations is that a great deal of floristic change is to be expected in the present context of global warming (Box 7.1). On the other, past climatic fluctuations offer one explanation for the disjunct distribution patterns often observed in bryophytes. For instance, the Pleistocene Period, which lasted from 1 640 000 to 10 000 yr BP, was marked by dramatic glaciations, which dissected previously broader, continuous ranges such as that of the liverwort genus *Nipponolejeunea* into disconnected, unglaciated refugia (Fig. 6.16).

7.2 Landscape ecology

At the landscape scale, community patterns tend to parallel variations in environmental conditions, such as geology, soil conditions and land use. Figure 7.4 illustrates the match between patterns of species richness in 87 4 × 4 km grid-squares in an area of Belgium with a discontinuous forest cover within an agricultural landscape. The positive relationship between species richness and forest cover clearly suggests that species distributions are controlled by the latter. For example, the distribution of the leafy liverwort

Box 7.1
Bryophytes and global warming

Because they are poikilohydric and thus receive water directly from ambient air humidity, bryophytes are, perhaps more than vascular plants, threatened by global climate change. The intensity of global warming resulting from the 'greenhouse effect' associated with increasing CO_2 concentrations in the atmosphere is predicted to vary along a gradient of latitude. Arctic and subarctic areas are likely to experience a particularly high increase in winter temperature varying from 2°C to as much as 14°C. As a consequence, high latitude ecosystems will be the first to be affected before the effects on other ecosystems become apparent (Wasley *et al.* 2006). In particular, the increase of soil temperatures will result in increased rates of evapotranspiration, a factor that is crucial for boreal peatlands.

According to the predictions of a bioclimatic model using temperature and precipitation as variables (Gignac *et al.* 1998), the southern borderline of boreal peatlands would be shifted 780 km northwards and would essentially disappear south of 60° latitude if CO_2 concentrations were to be doubled (Box 7.1 Fig. 1). Climate change is therefore likely to alter accumulation of carbon by *Sphagnum*, with consequences for the sink function of northern peatlands and, hence, have major feedback effects on climate itself (see Section 2.1).

The consequences of global warming on temperate ecosystems are expected to be less dramatic. Bates *et al.* (2005) found that the effects of simulated summer drought, summer supplemented rain (20% increase based on the previous 10-year average) and winter warming (3°C above ambient), were relatively modest in a calcareous grassland bryophyte community. Total bryophyte cover, but not species richness, was significantly affected by summer drought (Box 7.1 Fig. 2). The responses of individual species to imposed summer drought and/or winter warming, including 'increasers' (e.g. *Fissidens dubius* (Box 7.1 Fig. 3a)) and 'decreasers' (e.g. *Rhytidiadelphus squarrosus* and *Lophocolea bidentata* (Box 7.1 Fig. 3b)), were significant for 7 of the 13 species investigated. One of the reasons why the experimental manipulations exhibited only a limited impact is that many bryophytes are able to withstand repeated desiccation events without injury, typically resuming normal metabolism within minutes to a few hours of re-hydration (Section 8.1.1). In addition, sufficient hydration for normal metabolism might be attained by dewfall, which is perhaps the predominant source of moisture for grassland bryophytes.

Box 7.1 (cont.)

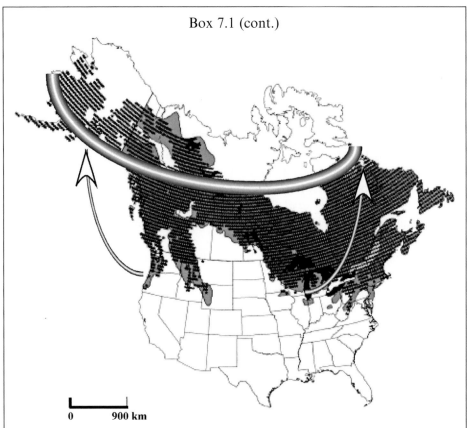

Box 7.1 Fig. 1. Relationship between the actual and estimated distributions of *Sphagnum*-dominated peatlands in North America. The light grey area corresponds to the observed distribution and the darker area indicates 0.5° latitude by longitude grid cells in which peatlands were estimated to be found by a predictive model employing mean annual temperature and mean annual total precipitation as predictors (reproduced from Gignac *et al.* 2000 with permission of Blackwell). Arrows and curve indicate the extent of the northwards migration of peatlands if CO_2 concentrations were to be doubled (after Gignac *et al.* 1998).

More spectacular impacts of global warming on bryophytes of temperate areas, including northward and eastward range extensions of Mediterranean species, have been reported (Frahm & Klaus 2001). For example, the conspicuous increase of the Mediterranean moss *Rhynchostegium megapolitanum*

Box 7.1 (cont.)

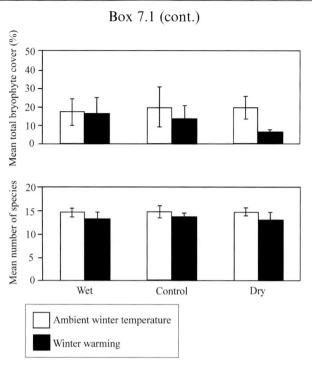

Box 7.1 Fig. 2. Mean total bryophyte cover (%) and mean species richness in a UK limestone grassland bryophyte community exposed to manipulated rainfall (summer drought (dry) and supplemented summer rainfall (wet)) and increased winter temperature. Summer drought was imposed in July and August by semitransparent rainshelters sliding over the plot automatically whenever rainfall was detected. Additional summer rainfall was supplied by watering at regular intervals from the beginning of June until the end of September to give a 20% increase based on the previous 10-year average for this period. Winter warming corresponds to an increase in soil surface temperature by 3°C above ambient by using electrical soil heating cables (reproduced from Bates *et al.* 2005 with permission of Blackwell).

in Central European dry grasslands correlates with a significant increase in temperature from 1921 onward (Zechmeister *et al.* 2007a). The extent to which the range expansion of such species is actually due to global warming remains, however, to be experimentally demonstrated.

Box 7.1 (cont.)

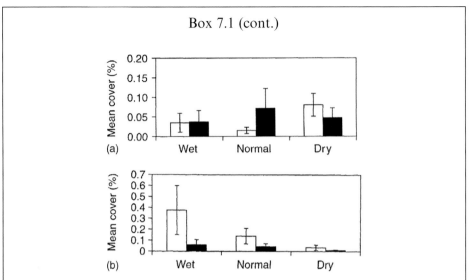

(a)

(b)

Box 7.1 Fig. 3. Mean cover of *Fissidens dubius* (a) and *Lophocolea bidentata* (b) in a limestone grassland bryophyte community exposed to manipulated rainfall (see Box 7.1 Fig. 2) (reproduced from Bates *et al.* 2005 with permission of Blackwell).

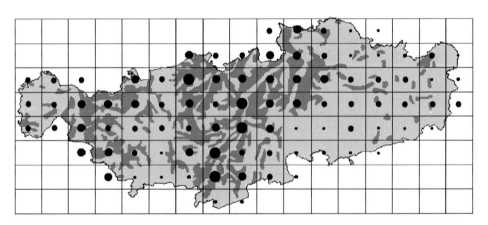

Number of species per 4x4 km

● Wooded area

- · 55–71
- • 72–88
- ● 89–104
- ● 129–182

4km

Fig. 7.4. Total number of bryophyte species recorded in 87 4 × 4 km grid-squares in Walloon Brabant (Belgium) superimposed on a background of forest cover (reproduced from Vanderpoorten & Engels 2003 with permission of Springer).

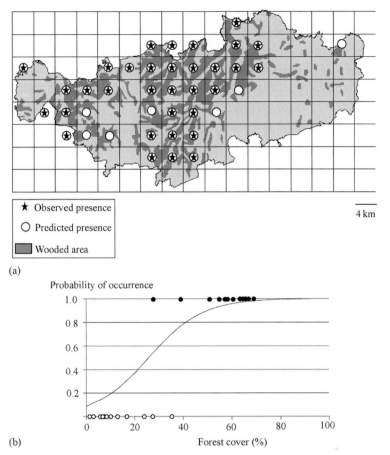

Fig. 7.5. Relationship between forest cover and the distribution of the leafy liverwort *Calypogeia fissa* in Walloon Brabant (Belgium). (a) Predicted and observed occurrence of *C. fissa* in each of the 87 4 × 4 km grid-squares of the area superimposed on a background of forest cover. (b) Probability of occurrence of *C. fissa* depending on forest cover in Walloon Brabant (redrawn from Vanderpoorten & Engels 2002). The curve is obtained by dividing forest cover into classes and taking the average frequency of the species in each class as an estimate of its probability of occurrence. Plain and open circles represent actual presences and absences of the species in each of the 87 grid-squares, respectively.

Calypogeia fissa can be easily superimposed on a background of forest cover (Fig. 7.5a), which can be used as a predictor of the occurrence of the species (Fig. 7.5b). The match between forest cover and species occurrence is, however, only partial. In certain grid-squares, forest cover is theoretically sufficient to allow for the presence of *C. fissa*, yet the species was not actually observed (white circles in Fig. 7.5a).

Aside from the failure to record a species in an area where it does actually occur, the mismatch between these observations and predictions might be explained by three main reasons. First, the occurrence of the species may be regulated by local environmental conditions. Second, species may interact with each other. Competition, in particular, is likely to exclude a species from an apparently favourable habitat. Third, local habitat conditions may have changed and the species may not exhibit the dispersal potential that would be necessary to colonize new spots. These three factors are examined in the remainder of this chapter.

7.3 Population ecology

7.3.1 Ecological niche

The presence of a species and its ability to reproduce are governed by a multitude of environmental factors, the combination of which defines its fundamental ecological niche. Measures of niche overlap evaluate the degree to which species subdivide the habitat space within the community and indicate the extent to which different species may utilize the same resources.

Epiphytic communities provide a classic example of niche differentiation. Epiphytes most often colonize the dead outer tissue (i.e. the bark) of tree trunks and branches without drawing water or food from living tissues. They may, however, profit from substances dissolved by rainwater from the dead outer bark of the host. Epiphytes also sometimes grow on herbs, on the surface of leaves, or even on other bryophytes, in which case they are termed epiphylls (Fig. 7.6).

The tree habitat is rather extreme. Its main features are a constant supply of new areas for colonization and a wide range of temperature and humidity conditions. When progressing towards the crown, wind force and light intensity increase and hence fluctuations in temperature and humidity (Fig. 7.7). The range of variation of these factors depends, however, on the local climatic conditions and on the density of the surrounding vegetation (Moe & Botnen 2000). Because of a constant supply of moisture, corticolous bryophyte communities of montane cloud forest experience, for instance, less drought stress and are more species rich than those of low montane and lowland forests (Acebey *et al.* 2003).

The composition and richness of corticolous bryophyte communities are further determined by the pH and nutrient content of the bark. The pH of the bark largely depends on the tree itself, ranging from acidic (e.g. in beech and birch) to neutral (e.g. in elms and elders). These differences in bark chemistry

Fig. 7.6. Epiphyllous *Frullania* sp. from the Azores (photo J.-P. Frahm).
See plate section for colour version.

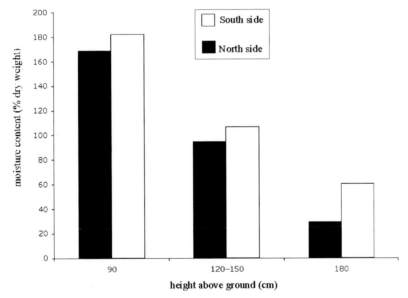

Fig. 7.7. *Liriodendron* bark moisture (% dry weight) (southeast face, 21
March 1937) in a northeast American woodland (data from Billings & Drew
1938).

can locally result in host preference (González-Mancebo *et al.* 2003, Szövényi
et al. 2004). Nutrient-rich, basic bark usually harbours a higher species
richness than acidic, nutrient-poor bark (Fig. 7.8). Along the trunk, bark
exudates, which are nutrient-rich and display a neutral or high pH, dissolve in
rainwater and accumulate at the base, creating a vertical ecological gradient.
Hence, the tree can be subdivided into several ecological units including the

Mean number of species

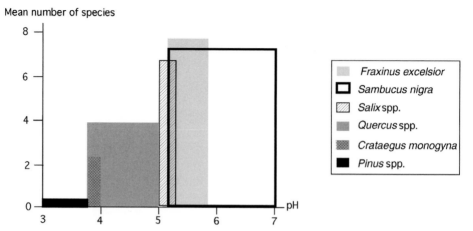

Fig. 7.8. Mean number of epiphytic bryophyte species recorded per tree species in southern England depending on the pH of the bark (data from Bates *et al.* 1997 for bryophyte species diversity and from Barkman 1958 for pH range values).

crown, crown base, middle part of the trunk, sub-median part and the base. Within each ecological unit, bark fissures generate a microtopography of prime importance in trees such as pines or eucalypts, wherein fissures are a more stable habitat than the projecting scales that are regularly exfoliated. Fissures provide a shady, moist habitat retaining snow and dust and are sheltered against wind. They also create wounds that produce exudates which are often of higher pH than bark itself.

Epiphytes therefore typically exhibit both a vertical and a horizontal zonation (Barkman 1958, Bates 1992, McCune *et al.* 1997, Williams & Sillett 2007). They segregate vertically from the base to the crown along gradients of humidity, pH and nutrient content. In the upper montane forest of Rwenzori (Uganda), for example, Pentecost (1998) described a zonation, wherein large cushion-forming mosses and hepatics are followed by small, creeping leafy liverworts on twigs when progressing from the tree base to the upper canopy (Fig. 7.9).

The horizontal zonation is mostly influenced by exposure. On elm trees of the coastal region of The Netherlands, Barkman (1958) described for example a horizontal zonation, wherein epiphytes are absent from the east side, which does not receive wet coastal winds; xerophytic species, such as *Syntrichia laevipila*, colonize the south side; and large mats of pleurocarpous mosses, including *Homalothecium sericeum, Brachythecium rutabulum* and *Oxyrrhynchium hians*, colonize moist, north and northwest sides (Fig. 7.10). Within each ecological unit, bark microtopography further generates a mosaic

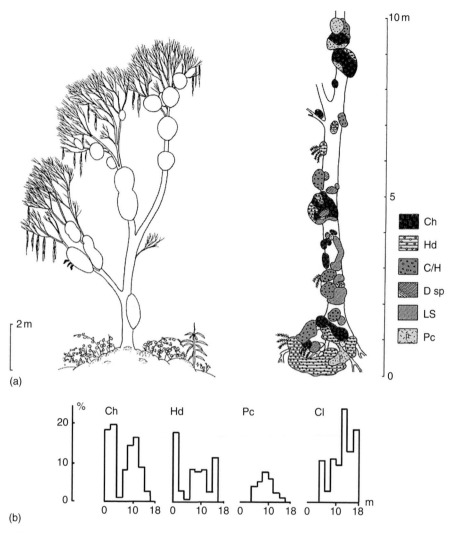

Fig. 7.9. Vertical zonation (a) and mean percentage cover (b) of the major epiphytes on *Philippia* in upper montane forest of Rwenzori (Uganda). Ch: *Chandonanthus hirtellus* subsp. *giganteus*; Hd: *Herbertus dicranus*; C/H: *Chandonanthus/Herbertus* mosaic; D: *Dicranum* sp.; Ls: *Lepidozia* cf. *stuhlmannii*; Pc: *Plagiochila colorans* and *P. ericifolia*; Cl: *Cheilolejeunea* sp. (reproduced from Pentecost 1998 with permission of Blackwell).

of microhabitats. For example, Barkman (1958) described the mosaic of species inhabiting beech bark in The Netherlands (Fig. 7.11). Wound exudates induce a vertical zonation of neutrophytic species, including *Orthotrichum diaphanum*, *Syntrichia laevipila* and *Zygodon viridissimus*, which are normally absent from acid beech bark. The last two species grow lower, presumably due to greater

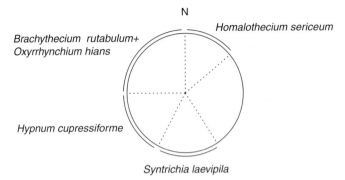

Fig. 7.10. Schematic horizontal variation in bryophyte epiphytes on a coastal elm tree in The Netherlands at tree base (redrawn from Barkman 1958).

moisture near the ground. In contrast, acidophilous species, such as *Lophocolea heterophylla*, develop far from the wound.

7.3.2 Species interactions

Different species thus tend to utilize different portions of the resource continuum available. Although niche breadth sometimes shows considerable variation for different species and different factors, and although some taxonomically closely related species can show much overlap for some factors, the niches of most species can often be separated on the basis of two or more factors. A species may, however, be excluded from parts of its fundamental niche because of, for example, historical factors and competition. This reduced niche is termed the realized niche. The competitive exclusion principle predicts that species avoid competition by occupying different niches, creating a spatial pattern that represents habitat partitioning corresponding to habitat heterogeneity (Slack 1990). However, a complete absence of niche differentiation among species has been reported in many cases. In some instances, absence of niche differentiation allows for the coexistence of different species within the same niche. In others, some species may exhibit a much greater ability to invade space than others from the same niche. These opposite trends regulate the fine spatial structure of bryophyte populations (Økland & Bakkestuen 2004).

7.3.2.1 Positive interactions

At low to moderate densities, growth is constrained by water availability. Moderately dense stands are dehydrated less rapidly than loose stands or

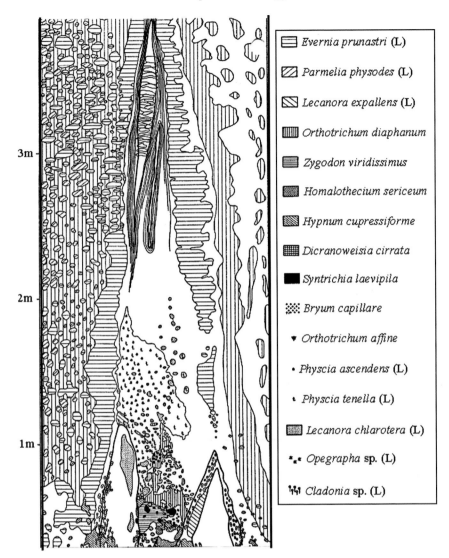

Fig. 7.11. Mosaic of cryptogamic vegetation comprising lichens (L) and bryophytes along the first 4 m of an old beech trunk in The Netherlands (redrawn from Barkman 1958).

isolated shoots because a dense packing of shoots may reduce water loss by effectively reducing the diameter of capillary spaces among close neighbours. Hence, shoot size, biomass production and species diversity are often positively related to carpet density (Fig. 7.12) (Økland & Økland 1996, Økland 2000, Bergamini *et al.* 2001a). Økland (1994) was the first to talk about 'favourability gradients' in bryophytes, whereby the water content of the shoots rises as their density increases. Bryophytes growing in dense communities

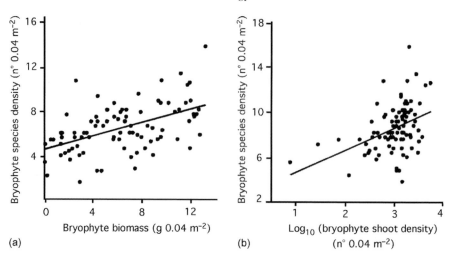

(a)   (b)

Fig. 7.12. Relationships between bryophyte species diversity and bryophyte biomass (a) and shoot density (b) in a montane wetland bryophyte community (reproduced from Bergamini *et al.* 2001a with permission of Blackwell).

are therefore able to remain physiologically active for a larger part of the growing season, resulting in greater biomass (Bates 1988).

The positive effect of shoot density on shoot size brings about a positive spatial structure in the community. That is, size difference among shoots and the risk of producing shoots with no offspring is expected to increase with increasing distance between shoots. This is because the mutually positive influence of neighbours makes all shoots in a neighbourhood larger and more likely to survive than if they occurred as isolated shoots.

7.3.2.2 Competitive exclusion

At high carpet density, however, many shoots are deprived of light and may suffer from growth reduction because they are overtopped by larger neighbours. Interactions among shoots then become competitive and a negative spatial structure will be observed. In this case, size difference among shoots and the risk of producing shoots with no offspring will decrease with increasing distance between shoots. Indeed, with increasing stand density, light deficiency will restrict the growth of an increasing number of overtopped shoots. The competition for light will create greater size variation among neighbouring shoots than among shoots separated by large distances.

Competition may occur at different stages of the life cycle. In *Tetraphis pellucida* for example, neither spores nor gemmae can become established in the presence of the gametophytes of other species (Kimmerer 1993). Similarly,

Cronberg *et al.* (2006b) attributed the lack of regeneration in established populations of the forest floor moss *Hylocomium splendens* to the fact that spores are unable to germinate or that sporelings are suppressed by mature plants. Establishment of spores may be hampered by the secretion of substances that prevent the growth of the protonema. Such substances are thought to occur in *Funaria hygrometrica* and *Polytrichum juniperinum* and appear to be non-species-specific, preventing all protonema in the vicinity from growing. In such a case, establishment is only possible in the complete absence of the parental gametophytes, that is, in newly disturbed areas (Kimmerer 1993, Sundberg & Rydin 2002, Cronberg *et al.* 2006b).

At the gametophytic stage, the existence of interspecific competition has been an area of controversy because many bryophyte communities are seemingly open, with large patches of appropriate habitat still available. An increasingly large body of literature suggests, however, that competitive exclusion does occur and may explain the predominance of one species over another within the same niche. For example, Frego and Carleton (1995a) observed that boreal forest ground is prominently covered by the moss *Pleurozium schreberi*, but failed to detect habitat partitioning with three other species of boreal forest ground on single or combined gradients of temperature, light and humidity. Figure 7.13 shows the frequency of occurrence of the four species along the same environmental gradients. All four species are able to occupy the same range of environmental conditions. Reciprocal transplant experiments confirmed that all species are capable of surviving in all microsites currently occupied by others in their community, at least over three growing seasons.

The niches for the four species are thus broad and overlapping, but the four species display differences in growth rates (Frego & Carleton 1995b). Growth, as measured by biomass gains after one year, is highest in *Ptilidium* and *Pleurozium*; intermediate in *Ptilium*; and slowest in *Dicranum* (Fig. 7.14). Hence, the predominance of *Pleurozium* in boreal forests does not result from its better adaptive potential to a range of ecological conditions, but rather from its ability to remove others through competitive interactions by means of a growth strategy oriented towards the fast occupancy of space.

Species indeed differ in their architecture and architectural plasticity, which has been described under the concept of 'growth forms' (Box 7.2). The space available to terrestrial bryophytes is upwardly dynamic, which means that previous occupants are not removed but overtopped, leaving the accumulated history of the community slowly decomposing below. Subtle differences in growth responses are critical to persistence within the community. Species with a rigid vertical structure and few branches (typically, acrocarpous

photosynthetically active radiation (μmol/m²/s) class

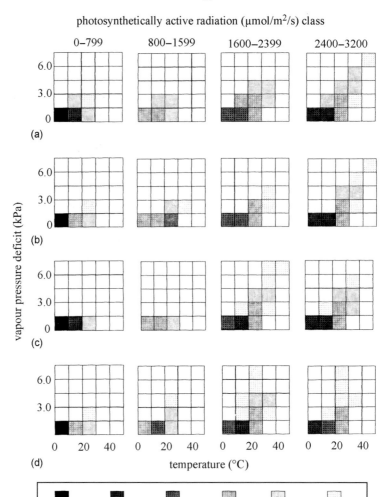

Fig. 7.13. Frequency distribution for temperature (°C), vapour pressure deficit (kPa) and photosynthetically active radiation ($mol\,m^{-2}\,s^{-1}$) in microsites containing *Dicranum polysetum* (a), *Ptilidium ciliare* (b), *Pleurozium schreberi* (c) and *Ptilium crista-castrensis* (d) in a Canadian boreal forest (reproduced from Frego & Carleton 1995a with permission of *Canadian Journal of Botany*).

species such as *Dicranum*) grow more slowly than the more frequently branched species. Slow elongation in *Dicranum* is associated with its rigid habit as well as its limited branching ability and shape plasticity. This structural form presumably requires more biomass, hence net production, to add to unit length. Such morphological features may represent a strategy that

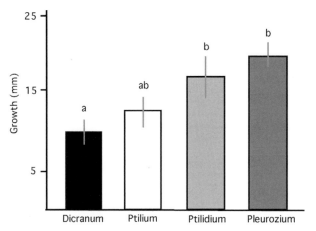

Fig. 7.14. Shoot elongation of four dominant bryophytes of the boreal forest ground in Canada during a reciprocal transplantation experiment, wherein three shoots per species were placed individually in each of four beds of each species, the conspecific bed serving as control. Each shoot was tagged with polyester threads and elongation above thread was measured after one year. Bars are means $\pm$ SE; those with the same letters are not statistically different at $p < 0.05$ (reproduced from Frego & Carleton 1995b with permission of the authors and *The Bryologist*).

Box 7.2
Bryophyte architectural growth forms

The increasing interest in vegetation texture and structure calls for a detailed system of plant growth forms. Growth forms can be defined as types of plants sharing the same gross morphology, i.e. the same architecture.

Bryophytes, despite their small size of a few millimetres to decimetres (up to a metre in rare instances), exhibit a vast array of morphologies that have been accommodated in three main categories depending on the direction of main shoots (Box 7.2 Fig. 1). These categories include 'turfs', which correspond to the erect habit of acrocarps; 'cushions', which correspond to plants radiating from a central point; and 'mats', which correspond to the creeping or ascending habit of pleurocarps. Each category can be further subdivided depending on features such as shoot size and density. Within the 'mat' growth form for example, 'wefts' are characterized by a loose intertwining of straggling shoots, which stick up and arch over each other in different directions; 'pendents' typically hang on branches; and thread-like mats are composed of very delicate-looking shoots with tiny leaves, forming a fine 'feathery' mat, or sometimes just single strands between other plants or on bare surfaces.

Box 7.2 (cont.)

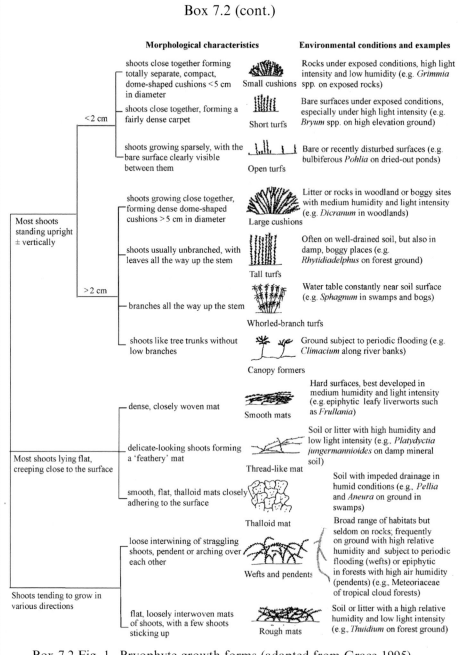

Morphological characteristics		Environmental conditions and examples

shoots close together forming totally separate, compact, dome-shaped cushions <5 cm in diameter — **Small cushions** — Rocks under exposed conditions, high light intensity and low humidity (e.g. *Grimmia* spp. on exposed rocks)

shoots close together, forming a fairly dense carpet — **Short turfs** — Bare surfaces under exposed conditions, especially under high light intensity (e.g. *Bryum* spp. on high elevation ground)

shoots growing sparsely, with the bare surface clearly visible between them — **Open turfs** — Bare or recently disturbed surfaces (e.g. bulbiferous *Pohlia* on dried-out ponds)

shoots growing close together, forming dense dome-shaped cushions >5 cm in diameter — **Large cushions** — Litter or rocks in woodland or boggy sites with medium humidity and light intensity (e.g. *Dicranum* in woodlands)

shoots usually unbranched, with leaves all the way up the stem — **Tall turfs** — Often on well-drained soil, but also in damp, boggy places (e.g. *Rhytidiadelphus* on forest ground)

branches all the way up the stem — **Whorled-branch turfs** — Water table constantly near soil surface (e.g. *Sphagnum* in swamps and bogs)

shoots like tree trunks without low branches — **Canopy formers** — Ground subject to periodic flooding (e.g. *Climacium* along river banks)

dense, closely woven mat — **Smooth mats** — Hard surfaces, best developed in medium humidity and light intensity (e.g. epiphytic leafy liverworts such as *Frullania*)

delicate-looking shoots forming a 'feathery' mat — **Thread-like mat** — Soil or litter with high humidity and low light intensity (e.g., *Platydyctia jungermannioides* on damp mineral soil)

smooth, flat, thalloid mats closely adhering to the surface — **Thalloid mat** — Soil with impeded drainage in humid conditions (e.g., *Pellia* and *Aneura* on ground in swamps)

loose interwining of straggling shoots, pendent or arching over each other — **Wefts and pendents** — Broad range of habitats but seldom on rocks; frequently on ground with high relative humidity and subject to periodic flooding (wefts) or epiphytic in forests with high air humidity (pendents) (e.g., Meteoriaceae of tropical cloud forests)

flat, loosely interwoven mats of shoots, with a few shoots sticking up — **Rough mats** — Soil or litter with a high relative humidity and low light intensity (e.g., *Thuidium* on forest ground)

Most shoots standing upright ± vertically (<2 cm / >2 cm)

Most shoots lying flat, creeping close to the surface

Shoots tending to grow in various directions

Box 7.2 Fig. 1. Bryophyte growth forms (adapted from Grace 1995).

Box 7.2 (cont.)

Growth forms are defined on the basis of plant architecture only. Yet, the distribution of each growth form over different habitats is, to some extent, constrained by competition and environmental conditions (During 1992, Grace 1995, Sillett *et al.* 1995, Bates 1998, Holz *et al.* 2002). In particular, the architectural features of canopies affect the water balance of bryophytes. Bryophytes therefore tend to exhibit similar adaptive convergence of forms depending on the level of water availability. Among African, Asian and South American tropical mountain epiphytes, Kürschner and Parolly (2005) observed, for example, that mats, wefts and pendent forms, which enhance photosynthetic activity whilst permitting rapid water uptake from water-saturated air, dominate ($>85\%$ of mean percentage cover) at the sub-montane and montane levels. In disturbed, sun-exposed areas and at higher elevation, cushions and turfs become progressively dominant. In fact, a dense packing of shoots is better adapted to xeric conditions than loose mats because the small space among close neighbours favours water retention by capillarity.

---

allows *Dicranum* to resist overtopping by other species, at the expense of lateral extension. By contrast, pleurocarps such as *Ptilium* and *Pleurozium* may be lax but branch diffusely, scrambling over plants below. *Pleurozium*'s slight but consistent advantage, compounded over decades, could account for its observed dominance. In addition, it is the most morphologically plastic, with forms ranging from thread-like single axes to robust sprawling multiple branches of several orders. The combination of variable branch size, branch frequency and lax form may also contribute to the species invasiveness, i.e. its ability to extend into adjacent spaces.

Two main strategies have been developed by weak competitors to avoid being outcompeted by other species. First, some species were shown to be able to shift their niche optimum towards a niche that is less favourable, but less prone to competition. For example, Wiklund & Rydin (2004b) studied the factors that influence growth of the epiphytic moss, *Neckera pennata*. They found that growth is positively influenced by a rather high position on the tree and occurrence on trees with low diameters and negatively influenced by the presence of other species. However, when the latter factor is removed, the effect of the two ecological factors becomes statistically non-significant, exactly as if *Neckera* were forced to grow higher on the tree by strong competitors.

The second strategy to avoid competition is to escape into newly created habitats. The moss *Tetraphis pellucida* is, for example, a weak competitor of decaying logs. The species relies on disturbance of its habitat, through, for example, brown-rot decay and mechanical disruption, to continually open patches of open space. When disturbance is experimentally reduced, competition with stronger competitors increases and *T. pellucida* becomes locally extinct. The species maintains ecological dominance by dispersing and exploiting frequently disturbed microsites (Kimmerer 1993). In this respect, *T. pellucida* is aided by its dual reproductive mode. In fact, *T. pellucida* produces both spores, which are mostly involved in long-distance dispersal, and vegetative propagules, which mostly support the growth of local populations. Establishment frequency is much higher for gemmae than spores, indicating different roles for these two types of propagules, that is, routine short-distance dispersal via gemmae and random long-distance dispersal by spores. This difference in dispersal range between sexual versus asexual diaspores is striking when one compares the dispersal range of *T. pellucida* and that of *Dicranum flagellare*, a species that almost entirely relies on asexual propagules for dispersal. *Dicranum flagellare* occurs more frequently on aggregated logs, with other substrata less than 50 cm away. Its frequency of occurrence drops significantly with increasing isolation (Fig. 7.15). In contrast, *T. pellucida* commonly establishes in gaps on previously uncolonized logs and its frequency of occurrence does not increase with log aggregation. The dual dispersal mode of *T. pellucida* thus enables the species to persist in established colonies, while allowing it to explore the landscape in search of new potential habitats when local conditions become unfavourable. Indeed, *T. pellucida* is capable of switching dispersal mode from the production of asexual diaspores to sexual reproduction depending on local population density. Everything happens as if a high population density triggered the production of long-distance dispersal units to colonize new habitats (Kimmerer 1994).

7.3.3 The necessity to move and adapt within a dynamic landscape: the trade-off of life strategies

One way to avoid exclusion by competition is thus to colonize new habitats. These habitats eventually become unsuitable, either because they are progressively invaded by stronger competitors, or because the habitats themselves vanish. In fact, many habitats are dynamic in the sense that they vanish at one place while being recreated somewhere else. For bryophytes, duration of habitats typically ranges from a few months or even weeks in the case of

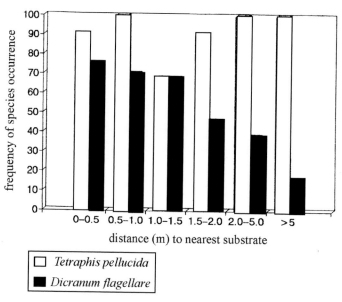

Fig. 7.15. Frequency of occurrence of *Tetraphis pellucida* and *Dicranum flagellare* on logs varying in degree of isolation by distance from nearest suitable substrate (reproduced from Kimmerer 1994 with permission of the author and *The Bryologist*).

pioneer species of temporary habitats such as dried-out ponds to thousands of years in the case of forest floors, bogs and rocky habitats (Fig. 7.16). The risk of extinction from these habitats, either due to competition, habitat longevity (Söderström & Herben 1997), or herbivory and parasitism (Davey & Currah 2006), constitutes a selective pressure to disperse and ensure population maintenance.

Bryophytes, like any other living organism, should thus ideally exhibit a great longevity through gametophytic adaptations to local environmental conditions (e.g. drought) and competition. At the same time, they should display a high reproductive potential, beginning at the earliest stage of their development and continuing throughout their existence, resulting in numerous offspring with high establishment rates. However, because the resources that are available for each individual are limited, some of these attributes become mutually exclusive. There is therefore a trade-off in the utilization of resources. The most favourable balance in the various trade-offs varies in relation to the nature and stability of the habitat, leading to the evolution of adaptive suites of life history features. In the classical theory of McArthur and Wilson (1967), organisms can be classified into two categories. 'r-species' produce many, small offspring with low chance of survival in

Table 7.1. *Bryophyte life strategies of During (1992)*

Strategy	Life span (years)	Number and size of spores	Reproductive effort	Typical habitat	Examples
Fugitive	<1	numerous, <20 μm	high	very temporary (e.g. burnt ground)	*Funaria hygrometrica*
Colonist	few	numerous, <20 μm	average	temporary (e.g. bare soil patches, branches)	pioneer species on ground (e.g. *Bryum* spp.), branches (e.g. *Orthotrichum* spp.), concrete (e.g. *Grimmia pulvinata*)
Perennial stayers	many	numerous, <20 μm	low	stable habitats (e.g. peat bogs)	*Polytrichum strictum*
Annual shuttle	<1	few, >20 μm	high	cyclic, temporary habitats (e.g. mud flats)	*Archidium alternifolium*
Dominants	many	few, >20 μm	low	open, stable habitats	mire *Sphagnum* species (e.g. *S. cuspidatum*, *S. lindbergii*)

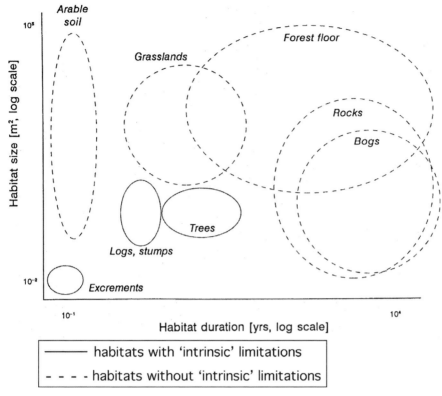

Fig. 7.16. Relationships between habitat size and duration of major habitats of temperate and boreal bryophytes (reproduced from Söderström & Herben 1997 with permission of Borntraeger).

early successional environments. By contrast, slow-growing, highly competitive 'K-species' reproduce after a long period of development and produce a few, large offspring with a high chance of survival in dense 'climax' environments.

A life strategy can thus be defined as a recurrent combination of life history traits that are predicted to occur prominently in response to particular ecological conditions. The r and K strategies defined above mostly apply to animals. In bryophytes, offspring size is not only related to the risk of high juvenile mortality but also to dispersal ability. Furthermore, reproduction does not specifically occur at a precise stage of the life cycle. For example, many bryophytes produce vegetative propagules and, more exceptionally in the case of neotenic species (Box 1.1), spores at the protonemal stage.

During (1979, 1992) showed how the concept of life strategies can be applied to bryophytes (Table 7.1). During's classification is based on the existence of two major trade-offs. The first of these trade-offs concerns the production of few, large spores or of many, small spores (Fig. 7.17). Large spores are thought to have a low dispersal capacity but better chances of successful establishment and a longer life span in the diaspore bank (Box 7.3). They are therefore prominently produced by shuttle species of unstable habitats that recur predictably at a given site. This is, for example, the case for hornworts in temperate areas, which are well adapted to regular disturbance in arable fields thanks to their diaspore bank (see Section 5.3), or of annual thalloid liverwort communities in xerotropical environments experiencing a severe drought season (see Section 3.3). Small spores, by contrast, are produced in such large numbers that many will reach distant sites. Such a strategy will be prominently adopted by colonist and fugitive species of ephemeral habitats. The classic example of a fugitive bryophyte is the weedy *Funaria hygrometrica* (Fig. 7.18), a species of habitats that occur unpredictably and are suitable for growth for a very short period of time such as, for example, burnt ground. The colonist strategy is comparable to the fugitive strategy, but colonists tend to occur in habitats (e.g. bare soil patches, branches, etc.) that are available for growth for somewhat longer periods. Colonists therefore have a potential life span of several years but substantially invest in reproductive effort by means of small spores and also often vegetative diaspores as their habitat eventually disappears (Fig. 7.19). Three categories of colonists can be recognized. Typical colonists, such as *Bryum dichotomum*, often produce both spores and gemmae and are typical of secondary successional communities such as ruderal habitats. Gap-dependent colonists, such as *B. rubens*, have short-lived gametophytes and assure population maintenance by subterranean tubers. They are typical on bare soils after disturbance of the ground vegetation (e.g. gaps in calcareous grasslands or forest floor caused by herbivores). Finally, pioneers disperse primarily by spores and colonize harsh environments (e.g. epiliths such as *Grimmia pulvinata* and *Tortula muralis* on concrete walls or pioneer epiphytes) during primary succession.

The second trade-off involved in During's classification concerns the potential life span of the gametophyte, which is negatively correlated with reproductive effort (Longton 1997). In the British moss flora, for example, all the short-lived fugitive species and annual shuttles commonly produce sporophytes; in colonists and long-lived shuttles, the proportion of species frequently producing sporophytes drops to about 60%, whilst this proportion is lower than 50% in perennial stayers and dominants (Fig. 7.20). Annual shuttle species avoid periods of environmental stress (e.g. severe

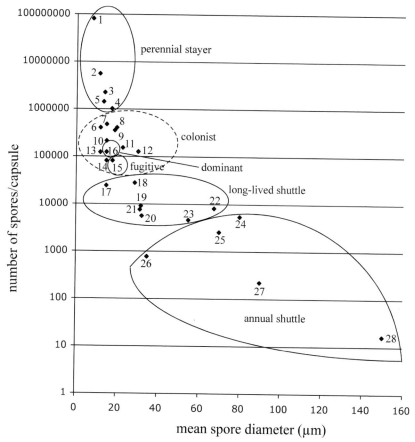

Fig. 7.17. Spore diameter and number of spores per capsule in a sample of bryophyte species with different life history strategies (data from Longton 1997). 1 = *Dawsonia lativaginata*; 2 = *Buxbaumia aphylla*; 3 = *Polytrichastrum formosum*; 4 = *Scapania undulata*; 5 = *Polytrichum piliferum*; 6 = *Diplophyllum albicans*; 7 = *Orthotrichum cupulatum*; 8 = *Funaria hygrometrica*; 9 = *Brachythecium rutabulum*; 10 = *Rhynchostegium confertum*; 11 = *Pohlia nutans*; 12 = *Mnium hornum*; 13 = *Ceratodon purpureus*; 14 = *Dicranella heteromalla*; 15 = *Leucobryum glaucum*; 16 = *Pleurozium schreberi*; 17 = *Lophocolea bidentata*; 18 = *Ptilidium pulcherrimum*; 19 = *Physcomitrium pyriforme*; 20 = *Phascum cuspidatum*; 21 = *Tortula truncata*; 22 = *Preissia quadrata*; 23 = *Pellia epiphylla*; 24 = *Conocephalum conicum*; 25 = *Reboulia hemisphaerica*; 26 = *Sphaerocarpos michelii*; 27 = *Riccia glauca*; 28 = *Archidium alternifolium*.

drought or frost) by producing short-lived gametophytes that vanish during these periods (Fig. 7.21). The resource allowance for the reproductive effort is typically high in such species, which tend to produce few, large spores that accumulate in soil pending a new period of favourable growth conditions.

Box 7.3
Ecological significance of bryophyte diaspore banks

Plants have to cope with unstable habitats in time (e.g. seasonal climate variations) and space (e.g. habitat degradation or destruction). When faced with the risk of local extinction, they may either disperse in an attempt to establish new populations or remain in the form of long-lived diaspores, from which new establishment will subsequently be possible under favourable growth conditions. Parts of these diaspores may become buried into the soil, requiring light for germination, constituting a bank of diaspores (During 1997). Because of the vulnerability of their gametophyte, bryophytes are, in particular, likely to rely more on stored propagules for their long-term survival than seed plants.

The diaspore bank of bryophytes allows species to survive unfavourable periods and facilitates rapid colonization after disturbance. The composition of the diaspore bank therefore influences the post-disturbance species composition and diversity (Jonsson 1993).

Not all species present in the actual vegetation are indeed represented in the diaspore bank. In boreal forests, for example, Jonsson (1993) found that typical colonizers of disturbed soil were very common in the diaspore bank, whereas some of the most abundant forest floor species were absent (Box 7.3 Fig. 1).

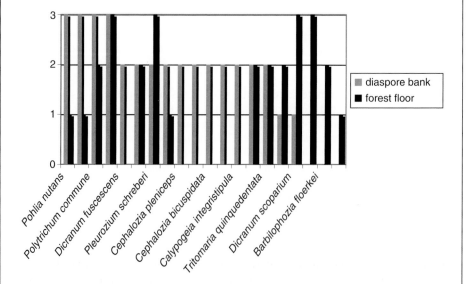

Box 7.3 Fig. 1. Differences in frequency of occurrence of the most common bryophyte species in the diaspore bank and forest floor in a Swedish boreal forest ($1 = <10\%$, $2 = 10$–50%, $3 = >50\%$) (data from Jonsson 1993).

Box 7.3 (cont.)

Similarly, During and ter Horst (1983) found that competitive pleurocarps, which dominated the above-ground bryophyte layer in a Dutch chalk grassland were seldom represented within the diaspore bank. By contrast, many acrocarpous species rarely found above ground composed the bulk of the bank. This discrepancy may reflect changes in the above-ground vegetation, but also differences in the life strategies: colonists, annuals and short-lived shuttle species are well represented, whereas perennials are rare.

Shuttle species typically produce a few, large spores that may be adapted to cyclic habitat conditions. In particular, seed dormancy, a well-known strategy of vascular plants to bridge unfavourable periods, has been recently demonstrated in bryophytes (Hock *et al.* 2004). In the annual shuttle *Phascum cuspidatum*, for example, the number of germinated spores increases strikingly after one year of storage, suggesting that some spores had been in a dormant phase (Box 7.3 Fig. 2). Bryophyte spores may thus be able to remain viable in the soil over decades (During 1997) and development of dormancy might facilitate this process.

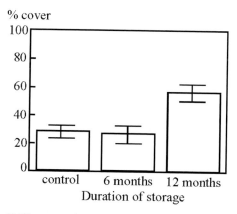

Box 7.3 Fig. 2. Differences in percentage cover of *Phascum cuspidatum* emerging from soil samples spread-out and kept in a greenhouse under appropriate growth conditions over 3.5 months immediately after soil sampling (control) and after 6 and 12 months of laboratory storage under air-dry conditions (reproduced from Hock *et al.* 2004 with permission of *Journal of Bryology*).

Diaspore banks reflect, to some extent, the composition of past rather than present vegetation. They accumulate genetic variability over generations and display more diversity than that present in the actual above-ground populations (Hock *et al.* 2008). Diaspore banks, therefore, play a key role in maintaining the high diversity of habitats that are naturally disturbed on a regular basis and in the restoration of degraded habitats.

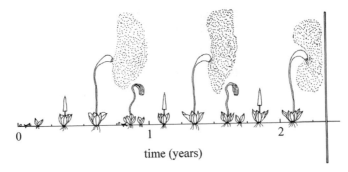

Fig. 7.18. The fugitive life strategy (During 1992). The vertical bar symbolizes the end of the period during which the habitat is suitable for the species. Reproduced from During (1979) with permission of *Lindbergia*.

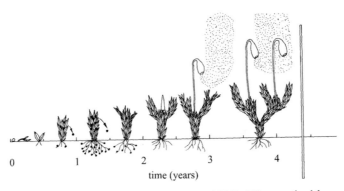

Fig. 7.19. The colonist life strategy (During 1992). The vertical bar symbolizes the end of the period during which the habitat is suitable for the species. Reproduced from During (1979) with permission of *Lindbergia*.

Often, annual species produce cleistocarpous capsules, wherein no specialized release mechanism exists and spores are directly included within the diaspore bank after the decay of the capsule. Cleistocarpy has evolved in parallel in several lineages of mosses (e.g. in the genera *Archidium* (Archidiaceae) and *Tortula* (Pottiaceae) and liverworts (all Sphaerocarpales and many Marchantiales). Annual shuttles are characteristic for cyclic habitats that disappear at varying rates but reappear at the same place (e.g. mud flats at the bottom of dried-out ponds and arable fields). Medium- and long-lived shuttles occur in somewhat longer-lasting micro-habitats. Typical examples of medium-lived shuttles include *Splachnum* species, which are restricted to dung and produce spores dispersed by flies over rather short distances (Box 4.2).

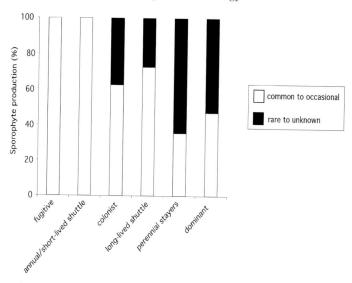

Fig. 7.20. Frequency (common to occasional versus rare to unknown) of sporophyte production within each of the six classes of life history strategies defined by During (1992) in the British moss flora (data from Longton 1997).

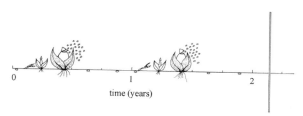

Fig. 7.21. The annual shuttle life strategy (During 1992). The vertical bar symbolizes the end of the period during which the habitat is suitable for the species. Reproduced from During (1979) with permission of *Lindbergia*.

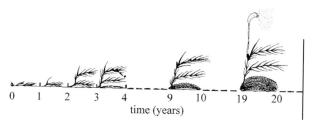

Fig. 7.22. The perennial life strategy (During 1992). The vertical bar symbolizes the end of the period during which the habitat is suitable for the species. Reproduced from During (1979) with permission of *Lindbergia*.

Conversely, stress-tolerant species invest much in gametophytic development, which enables them to survive periods of stress, and less in reproduction. Stress-tolerant species typically occur in somewhat permanent habitats (e.g. grasslands, peat bogs and forest floor), where they may survive for centuries (Bates 1989, Rydin 2009). Stress-tolerant species can be further divided into dominants and perennials. Dominants, which are only represented by a few *Sphagnum* species among bryophytes, produce few, large spores. Perennials, by contrast to dominants, produce small spores that serve for the occasional establishment of new populations (Fig. 7.22). Perennials can be slow-growing in dense vegetation such as peat bogs (e.g. *Polytrichum strictum*) or more aggressive, more competitive and capable of rapid, opportunistic growth (e.g. large pleurocarps such as *Brachythecium rutabulum*).

8

Physiological ecology

The correlations between species distribution patterns and environmental factors that were analyzed in the previous chapters at different spatial scales are a vital starting point in making hypotheses about causes and effects of biotic interactions, historical factors, reproductive characteristics and immediate physiological requirements. It appears that most bryophyte species tend to occupy well-defined niches, as if they had evolved discrete physiological optima for factors such as humidity, light, temperature and nature of the substratum. Water availability, in particular, seems to play a major role in bryophyte distributions. In fact, bryophytes are believed to have colonized the land from a fresh water origin (Mishler & Churchill 1985). This major ecological shift must have required the evolution of adaptive mechanisms to survive such harsh drying treatments in order to have successfully exploited terrestrial habitats (Oliver *et al.* 2005).

In the present chapter, we describe adaptive strategies to drought and show how water relations interact with other physiological attributes and constraints to shape the ecological and geographical distribution of bryophytes.

8.1 Water relations

8.1.1 Adaptive strategies to drought

All the basic metabolic processes of life take place only in the aqueous medium of a hydrated protoplasm. Drying to equilibrium with even moderately dry air is, hence, instantly lethal to most animals and plants. As a consequence, water availability is one of the most important ecological issues and evolutionary pressures on terrestrial life (Alpert 2005). Land plants have evolved three main strategies in response to drought, namely drought escape, drought avoidance and desiccation tolerance (Levitt 1980).

Drought escape refers to the ability to escape periods of drought, especially during the most sensitive periods of development. This involves the adoption of a life strategy that adapts the life cycle so that the plant dies during drought periods while desiccation-tolerant diaspores, including spores and vegetative propagules, remain viable in the ground. This strategy is best illustrated by species with an annual life cycle. Shuttle species (Section 7.3.3) are adapted to climates with a severe dry season by completing their life cycle within a few months during the wet season. Shuttle species avoid the dry season by producing large spores that have the ability to remain in the diaspore bank until above-ground conditions become favourable for growth again. This is the strategy adopted by annual mosses (e.g. in the Funariaceae and Pottiaceae) and thalloid liverworts (e.g. *Riccia*; see Section 3.3) in climates with a well-marked dry season, such as the Mediterranean (Kürschner & Parolly 1999, Kürschner 2003).

Drought avoidance (also commonly referred to as drought resistance) involves plants withstanding a period of drought by maintaining a favourable internal water balance. Avoidance relies on the maintenance of a chronic disequilibrium between the water content of the cell and the outer atmosphere through water retention. Avoidance is largely the privilege of vascular plants. In the latter, avoidance is aided by a series of morphological features that help to control air exchange (e.g. stomata) and limit evaporation (e.g. a waterproof cuticle). Water is pumped up from the soil by the roots and translocated through the plant by specialized conducting tissues. In bryophytes, this strategy is rare. It is linked to a series of morphological transformations to retain and transport water. The dense shoot packing and the presence of dead hyaline cells holding large amounts of water in leaves and around stems and branches in genera such as *Sphagnum* and *Leucobryum* are, for example, typically suggestive of drought resistance. Water conduction in the most sophisticated resistant species is internal. In so-called endohydric species, internal water conduction is facilitated by the presence of specialized water-conducting cells, such as fibrillose and porous hyalocysts in *Sphagnum* and some Dicranaceae (Section 4.1.1), hydroids in some mosses (Section 1.4.2) and perforated cells in some liverworts (Section 1.4.2). The level of resistance to drought displayed by bryophytes is, however, comparatively low with regard to that exhibited by vascular plants. Even in endohydric species, external conduction plays a significant role in water uptake. In fact, most bryophytes have evolved an alternative strategy to drought avoidance, namely desiccation tolerance.

Desiccation tolerance refers to the ability to dry to equilibrium with air that is moderately to extremely dry and then resume normal metabolic activity after rehydration. Alpert (2005) proposed that tolerance can be operationally defined as the ability to survive desiccation to a water content of 10%, i.e. of 0.1 g of water per gramme of dry weight. This corresponds to the point at which too little water remains to surround intracellular membranes and macromolecules and, therefore, to support metabolism. Desiccation tolerance thus requires that the plants reversibly cease metabolism during drought periods and develop adaptations to desiccation at the cellular level (Alpert & Oliver 2002).

Bryophytes are poikilohydric, which means that their water content is directly regulated by the ambient humidity. Most species are termed ecto-hydric because they take up water through the whole surface of the plant. The fact, that bryophytes are poikilohydric, means that it is not necessary for them to develop a root system to draw water from the soil. This alternative strategy enables them to grow on very hard surfaces such as rocks and tree trunks that are inhospitable to most vascular plants. Except in a few cases, for example, certain epiphytic Bromeliaceae, the vast majority of the latter are indeed unable to absorb water from their aerial organs and must draw it from soil. However, the poikilohydric condition brings about a major limitation to growth during dry periods. For example, Wiklund and Rydin (2004b) recorded the growth of a colony of the epiphytic moss *Neckera pennata* over four years and clearly demonstrated the prime importance of precipitation on colony development (Fig. 8.1). Bryophytes desiccate at the same time as or shortly after their substratum. Physiological activity and, hence, growth is restricted to periods of hydration, with the plant entering dormancy upon desiccation. The impact of water availability on growth is most pronounced in arid areas. For example, patches of the moss *Crossidium crassinerve* in the Mojave Desert experienced complete hydration only 8% of the time over a four-year period (Stark 2005). Since respiration strongly increases following desiccation, whereas photosynthesis recovers at a much slower rate (Alpert & Oechel 1985), one explanation for the relative scarcity of mosses in hot deserts is that a positive carbon balance, i.e. a positive net photosynthesis, is difficult or impossible to achieve in areas where moss patches are hydrated for insufficient periods of time.

In the absence of roots and a highly efficient internal water transport system, most bryophytes depend primarily on atmospheric water to sustain their needs. Water must be absorbed over much of their body surface, which must thus be permeable. The cuticle that seals the vascular plant body is most often reduced or even lacking on the vegetative body of bryophytes and stomata, which occur in the capsule wall of mosses and hornworts, are completely lacking from the leafy stem or the thallus.

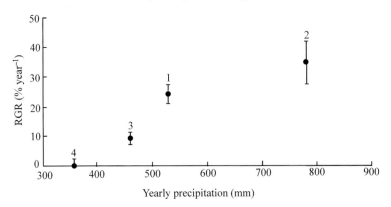

Fig. 8.1. Annual relative growth rate (RGR) of the epiphytic moss *Neckera pennata* over four years in relation to precipitation in a Swedish boreo-nemoral forest (reproduced from Wiklund & Rydin 2004b with permission of the authors and *The Bryologist*).

8.1.2 *Degrees of desiccation tolerance among bryophytes*

Many, and perhaps most, bryophytes can recover a normal metabolism upon rehydration after losing a substantial part of their cell water content (Proctor 2000). Species, however, vary greatly in the degree of desiccation they can endure. *Sphagnum* and tropical cloud forest *Leucobryum*, which occur in habitats that are unlikely to experience long-term water deficit, form compact tufts and exhibit anatomical adaptations for water storage. They seem better adapted for short-term drought resistance rather than desiccation tolerance. In fact, they can survive desiccation for only a few hours or a few days. By contrast, *Syntrichia caninervis*, a desert species, can remain at around -540 MPa (equilibrated to the atmosphere above silica gel at 2–4% relative humidity) for up to six years and still recover normal activity and growth upon rehydration (Oliver *et al.* 2005). As a most extreme example, thalli of the liverworts *Riccia* were found to be viable after 23 years in a herbarium (Alpert 2000).

Desiccation-tolerant mosses of open, sun-exposed habitats such as *Grimmia pulvinata* commonly tolerate severe negative water potentials of -300 MPa or less (Proctor 2001). In *G. pulvinata*, plants exposed to drought for periods of less than 60 days readily recover normal metabolic activity within a few minutes to hours, regardless of the initial intensity of desiccation (Fig. 8.2). Following prolonged drought exposure (120 to 240 days), recovery clearly proceeds most rapidly in the material desiccated at 20% relative humidity. Plants kept at 20% relative humidity almost completely recover their photosynthetic activity after two days of rehydration, whereas recovery is

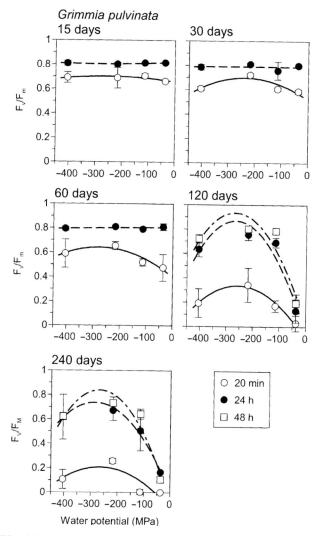

Fig. 8.2. Physiological response to increasing levels of desiccation (as measured by the water potential within the glass desiccator) during 15, 30, 60, 120 and 240 days in the xerophytic moss *Grimmia pulvinata*. The physiological response was measured as the intensity of damage to the photosynthetic apparatus (as assessed by the F_v/F_m ratio) 20 minutes, 24 and 48 hours after re-wetting of the specimens (reproduced from Proctor 2003 with permission of *Journal of Bryology*).

increasingly poor in plants kept at higher relative humidities (Fig. 8.2). By contrast, forest environments do not provide the selection pressure necessary for the evolution of adaptive strategies to survive such levels of stress. In the tropical cloud forest moss *Papillaria nigrescens*, for example, the degree of

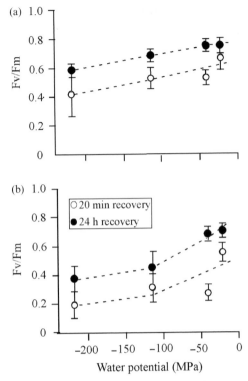

Fig. 8.3. Physiological response to increasing levels of desiccation (as measured by the water potential within the glass desiccator) during 6 (a) and 12 (b) days in the tropical cloud forest moss *Papillaria nigrescens*. The physiological response was measured as the intensity of damage to the photosynthetic apparatus (as assessed by the F_v/F_m ratio) 20 minutes and 24 hours after re-wetting of the specimens (reproduced from Léon-Vargas *et al.* 2006 with permission of Blackwell).

recovery is negatively correlated with increasing desiccation intensity (Léon-Vargas *et al.* 2006) (Fig. 8.3).

The difference between the pattern displayed by *G. pulvinata* on the one hand and *P. nigrescens* on the other parallels the difference between 'orthodox' and 'recalcitrant' seeds in angiosperms. Orthodox seeds display an optimum storage duration at high desiccation levels, preventing respiration and any potentially damaging micro-organismal activity, thus keeping the seed in a dormant state. By contrast, recalcitrant seeds lose their viability upon drying. It must be emphasized, however, that while recalcitrant seeds are strictly desiccation-sensitive, many bryophytes, including those of relatively moist forest environments, such as pendent pleurocarps of tropical rainforests, are able to tolerate days of drought at intensities as low as −40 MPa, at which most angiosperms do not survive (Proctor 2004).

8.1.3 Physiological adaptations to desiccation tolerance

From a physiological point of view, the adaptation to desiccation requires a cell structure that can lose most of its water without disruption and membranes that retain the essentials of their structure in the dry state or are readily and quickly reconstituted upon remoistening. All the essential metabolic systems of the plant (i.e. photosynthesis, respiration, etc.) must remain intact or be readily reconstituted. Desiccation-tolerant bryophytes survive desiccation by displaying both protective mechanisms during drought and repair mechanisms upon rehydration (Alpert & Oliver 2002).

8.1.3.1 Protective mechanisms involved in desiccation

Much of what is known about the cellular protection mechanisms involved in desiccation tolerance initially came from studies on orthodox seeds (Alpert & Oliver 2002). In the latter, a series of proteins, called Late Embryogenesis Abundant (LEA) proteins, accumulate at the same time as the acquisition of desiccation tolerance in the developing embryo. LEA proteins have also been identified in the tissues of all the desiccation-tolerant plants investigated so far. LEA proteins may protect the cellular components through a variety of mechanisms. LEA proteins sequester ions. This may be crucial as the increasing ionic concentrations during drying could cause irreversible damage to cellular proteins and structural components. The high affinity of LEA proteins for ions may also serve to trap valuable ions that would otherwise be lost by leaking through damaged membranes upon rehydration. Maintaining a high ionic concentration may, in turn, be essential for preserving a high water potential that is necessary to draw in water when it becomes available again. Finally, LEA proteins may also act as anchors in a structural network that stabilizes cytoplasmic components during drying and in the dry state. Other proteins, whose transcripts accumulate during dehydration in desiccation-tolerant tissues, have also been identified, but little has yet been done to confirm their role in desiccation tolerance.

In addition to protective proteins, desiccation-tolerant tissues are also characterized by the accumulation of soluble sugars. Sucrose is the only free sugar available for cellular protection in desiccation-tolerant mosses. The amount of this sugar in gametophytic cells of the xerophytic moss *Syntrichia ruralis* is approximately 10% of dry mass, which is sufficient to offer membrane protection during drying. Some sugars can replace water at hydrogen-bonding sites to preserve the native protein structure and spacing between phospholipids (Buitink & Leprince 2004). Sucrose is also involved in the process of vitrification, which is the formation of a biological glass. A glass is a liquid with an extremely high viscosity, which represents a semi-equilibrium, amorphous state between the solid and liquid states (Buitink & Leprince 2004). The shift from one state to another depends on the composition of the glass, the temperature and the water content. As the temperature and cell water content decrease, sugars inevitably turn into the amorphous state, which they reach at room temperature when the water content drops below approximately 10% dry weight (Hoekstra 2005). However, biological glasses exhibit different properties than if sucrose alone was involved. In particular, if the intracellular glass was composed of sucrose alone, a small increase in temperature would result in the collapse, crystallization and loss of macromolecular function and integrity. In fact, other cell solutes (for example, inorganic salts such as $MgCl_2$) and LEA proteins compose biological glasses and stabilize the latter by increasing the temperature of crystallization. Due to their high viscosity, biological glasses drastically reduce molecular movements, impede the diffusion of reactive compounds in the cell and prevent the cytoplasm from crystallizing. In short, glasses thus basically take the place of water in maintaining the structural and functional integrity of membranes.

8.1.3.2 Repair mechanisms upon rehydration

The level of protection that bryophytes are able to maintain is not completely sufficient to prevent damage, especially to membranes, during rehydration. Desiccation-tolerant bryophytes thus heavily rely on repair mechanisms during the first stages of hydration (Oliver 1991). Two main families of proteins are involved. Hydrins, 25 of which have been isolated in *S. ruralis*, are synthesized or substantially increase in concentration during the first two hours following rewetting. They are succeeded by rehydrins, which include 74 different molecules in *S. ruralis*. The mechanisms by which these molecules operate are still under scrutiny.

8.1.4 Why did vascular plants and bryophytes adopt contrasting adaptive strategies to drought?

8.1.4.1 Main functional differences between bryophytes and vascular plants

Bryophytes and vascular plants exhibit very different adaptive strategies to drought. Vascular plants adapted to the erratic availability of water on land by evolving a conducting system from underground roots to the above-ground leaves, regulating water loss by the stomata in the cuticularized epidermis (resistance to drought). Truly desiccation-tolerant 'resurrection' species are the rare exception, making up less than 0.15% of vascular plants. By contrast, bryophytes have retained the alternative strategy of photosynthesizing and growing actively when water is available and drying out and suspending metabolism when it is not (desiccation tolerance).

Because bryophytes are considered the closest relatives of the ancestors to these earliest terrestrial plants (Chapter 1), and because it is a widespread, but rare strategy among the earliest lineages in a vast array of organisms (Alpert 2006), desiccation tolerance could, at first sight, be seen as the most primitive strategy to conquer land. As plants evolved to fill the various niches available to them on land, the internalization of water relations would have reduced selection for tolerance. In this scenario, genes that had originally evolved for vegetative cellular protection and repair would possibly have been recruited for the desiccation tolerance of seeds (Oliver *et al.* 2000).

Recent phylogenetic developments suggest, however, that desiccation tolerance cannot be unambiguously designated as the most primitive strategy developed by land plants (Oliver *et al.* 2005). Rather, it seems that desiccation tolerance is the most appropriate, if not the only, adaptive strategy to drought for bryophytes. In particular, scale appears to be one of the major reasons for the differences in adaptive strategies to drought between vascular plants and bryophytes (Proctor *et al.* 2007). Large and small organisms are subject to quite different selection pressures, so that scale has profound effects on adaptive strategies. Bryophytes are not, and cannot be, simply scaled-down versions of vascular plants. For large organisms, gravity is a major force and surface tension is negligible. For a bryophyte, by contrast, surface tension is among the most powerful of forces and gravity is relatively trivial. As a result, many features of shoot and leaf architecture in bryophytes act through surface tension to store and to regulate the distribution and movement of extracellular water. For example, the presence of a dense tomentum around the stem of many unrelated moss lineages characteristic of bog habitats can be interpreted as a convergent, adaptive feature facilitating water uptake from the nearby water table through capillarity (Fig. 8.4). Gravity, by contrast, prevents water conduction by simple capillarity in taller plants, forcing them to develop a sophisticated internal conducting system. In the meantime, the size of the roots that would be necessary for endohydric bryophytes to pump up water from the water table, as do vascular plants, would make the proportions of the plants with such a small above-ground portion inviable. The small size of bryophytes also means that they have high ratios of surface area to volume, which in turn means that they dry out quickly. Therefore, at least a degree of desiccation tolerance is essential to them in any habitat where they are not kept constantly moist.

Fig. 8.4. *Breutelia azorica*, an Azorean endemic moss with a dense, brownish tomentum around the stem, facilitating water uptake and external trans-location by capillarity (photo J.-P. Frahm). See plate section for colour version.

8.1.4.2 Specificities of the physiology of desiccation tolerance in bryophytes

Even when vascular plants adopt a desiccation tolerance strategy, which is an extremely rare attribute, they display important differences to bryophytes. Tolerance is indeed largely environmentally induced in the former, whereas it is mostly constitutive in the latter. Desiccation-tolerant angiosperms employ an inducible cellular protection mechanism, which often relies, in part, on a drying-induced accumulation of abscisic acid (ABA). ABA triggers the accumulation of gene products that mediate the establishment of a cellular protection system prior to desiccation. Desiccation-tolerant angiosperms therefore cannot survive desiccation if water loss occurs rapidly (less than 12 h) and protein synthesis is required to establish tolerance (Proctor *et al.* 2007).

In bryophytes, by contrast, tolerance to rapid (within an hour) desiccation events suggests that cellular protection mechanisms are constitutive. This does not mean that desiccation tolerance is a fixed attribute in bryophytes

(Proctor *et al.* 2007). Tolerance by many species increases following a period of exposure to 96% relative humidity, and rapid drying (< 1 h) causes greater damage than slow drying (~ 3 h), probably because slow drying allows time for the setting in place of protective mechanisms (Proctor *et al.* 2007). Tolerance may, in addition, vary seasonally (Proctor *et al.* 2007). The point is, however, that no novel transcripts are recruited for protein synthesis in *S. ruralis* during drying since LEA proteins are continuously synthesized in bryophytes. Furthermore, the concentration of free sugars is always high and does not increase further upon drying.

The differences in the mechanisms of tolerance between these two groups reflect their biology. Desiccation-tolerant angiosperms generally have more effective morphological and physiological mechanisms to retard the loss of water. Everything happens as if the mechanism of desiccation tolerance that has evolved in these plants had taken advantage of these adaptations by being inducible. In other words, as the rate of water loss is relatively slow, there is time to establish the protective measures required. In bryophytes, by contrast, the ability to retain water is shaped by a trade-off between the necessity of permeability imposed by a mostly external water uptake and conduction and a trend for waterproofing in order to reduce evaporation. Hence, poikilohydry requires that the internal and external water potential are rapidly equilibrated. Bryophytes have therefore insufficient time to induce and set in place protective measures. The constitutive protection mechanism appears to be particularly efficient in preventing damage to the photosynthetic apparatus, as evidenced by the very rapid recovery of photosynthesis. In the meantime, it is most likely that desiccation tolerance, especially when constitutive, incurs a substantial metabolic cost that constrains growth rates. As a result, desiccation-tolerant species may be restricted to xeric habitats because they cannot compete with non-tolerant species with higher growth rates in mesic habitats (Stark 2005, Alpert 2006).

8.2 Light

Desiccation tolerance implies that bryophytes become metabolically inactive when dry. In some species, the leaves (e.g. the water-sacs of some leafy liverworts; see Section 3.1.1) or the growth form (e.g. dense cushions; see Box 7.2) are modified to prolong hydration. As necessary as water is for photosynthesis, excess water may limit carbon fixation as water-saturated tissues experience a high resistance to the diffusion of gaseous CO_2, the actual source of carbon for photosynthesis. In the moss *Syntrichia ruralis* for example, photosynthesis increases nearly four-fold upon loss of 50–65%

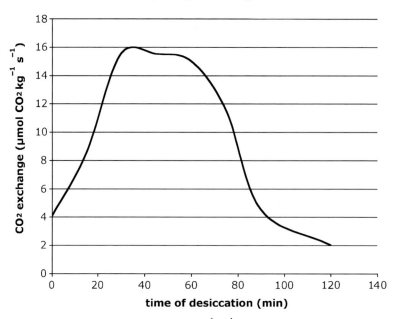

Fig. 8.5. CO_2 exchange ($\mu mol\, CO_2\, kg^{-1}\, s^{-1}$) during drying-out at 70% of relative humidity and 23°C in shoots of the moss *Syntrichia ruralis* that were initially oversaturated with water (redrawn from Tuba *et al.* 1996).

water from initially saturated samples dried slowly (Tuba *et al.* 1996; Fig. 8.5). These conflicting trends generate a trade-off between the necessity to avoid desiccation on the one hand and maximize the surface exposed for photosynthesis on the other. This trade-off is well illustrated in pendent pleurocarps, wherein capillary spaces between the permeable inner surface of the leaves and the stem hold external water, whereas the outer surface of the leaf is left dry to exchange gases (Fig. 8.6). Many species therefore carry at least a thin cuticle to keep a dry leaf surface that is available to absorb CO_2. In some cases, conspicuous granular or crystalline epicuticular waxes are present and give a glaucous, shiny appearance (Fig. 8.7). Similarly, in species with papillose leaf cell walls, the papillae line up water-conducting channels, while their apex remains dry and available for CO_2 diffusion.

The trade-off between the necessity of achieving a certain water content to permit metabolic activity and exposing a broad enough surface to facilitate gas exchange and light capture leads to photosynthesis mostly taking place during rainy or cloudy weather, when irradiance may often be less than 20% of that under full sun (Marschall & Proctor 2004). Hence, taxa of intensively insolated habitats, such as *S. ruralis*, are functionally 'shade' plants like the many bryophytes restricted to poorly illuminated niches. Bryophytes must thus be physiologically adapted to achieve a positive net photosynthesis at

Fig. 8.6. The pendent pleurocarpous moss *Garovaglia* forms massive curtains hanging on branches in the lowland rainforest of Danum Valley (Borneo). Note the concave leaves for external water storage on the dorsal side and the outer, convex surface of the leaf left dry to permit gas exchange (photo A. Vanderpoorten). See plate section for colour version.

low light levels, that is, exhibit low compensation points. In species of deep ravines and rock caves, such as *Fissidens serrulatus*, a compensation point as low as $7 \, \mu\text{mol photon m}^{-2}\text{s}^{-1}$ allows for a positive carbon balance in deep shade (Fig. 8.8). Photosynthesis in such species is, by contrast, inefficient at high light levels. Bryophytes of dense forests typically display saturation below $30 \, \mu\text{mol photon m}^{-2}\text{s}^{-1}$ (Gabriel & Bates 2003). In mosses in general, most species approach light saturation at about $600 \, \mu\text{mol photon m}^{-2}\text{s}^{-1}$, which corresponds to one third of full sunlight. In an experiment involving 39 mosses and 16 liverworts, Marschall and Proctor (2004) found that only three species exhibited light saturation above $1000 \, \mu\text{mol photon m}^{-2}\text{s}^{-1}$, which roughly corresponds to half of the highest intensity under midday

Fig. 8.7. *Pogonatum urnigerum* is a moss with a glaucous, shiny appearance due to the presence of waterproof crystalline epicuticular wax to keep leaf surfaces sufficiently dry for photosynthetic gas exchange, as evidenced by the drops of water on the leaf surface (photo J.-P. Frahm). See plate section for colour version.

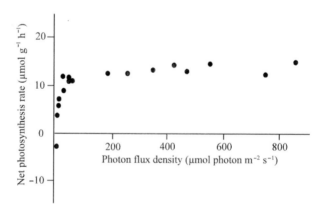

Fig. 8.8. Net photosynthesis of the laurel forest moss *Fissidens serrulatus* at 20°C as a function of light intensity (measured as photon flux density) (reproduced from Gabriel & Bates 2003 with permission of *Journal of Bryology*).

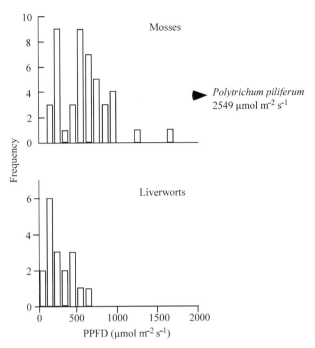

Fig. 8.9. Frequency distribution of 95% saturation values of photosynthesis activity (measured as photosynthetic electron transport (PPFD)) in a sample of 39 moss and 16 liverwort species (reproduced from Marschall & Proctor 2004 with permission of Oxford University Press).

summer sun (Fig. 8.9). As a comparison, light saturation is reached at 2125μ mol photon m^{-2} s^{-1} in the domestic apple, *Malus domestica*.

The highest saturation levels among bryophytes are reached in plants of open, wet habitats, with a guaranteed and constant supply of water. In the vast majority of the Polytrichaceae for example, the leaves are ornamented with green lamellae on the upper side of the costa (Section 4.1.1) that provide an area for CO_2 uptake, which is often more than six times the projected area of the leaf. Polytrichaceae thus appear fundamentally better equipped to photosynthesize in bright light than mosses with unistratose leaves. In Polytrichaceae, light saturation values are often $> 1000 \mu$ mol photon m^{-2} s^{-1}, making them competitive with small vascular plants in open habitats such as late-snow areas and peat bogs (Proctor 2005). However, the performance of the photosynthetic apparatus of some species seems to have evolved locally depending on water availability. For example, plants of *Sanionia* from dry and wet sites have adapted to low-light and high-light conditions by exhibiting contrasting photosynthetic response curves that reveal saturation at 500 and 1000μ mol photon m^{-2} s^{-1}, respectively (Ueno *et al.* 2006) (Fig. 8.10).

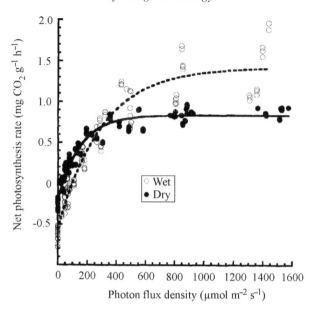

Fig. 8.10. Photosynthetic response to light intensity (measured as photon flux density) in the moss *Sanionia uncinata* from dry and wet habitats (reproduced from Ueno *et al.* 2006 with permission of *Journal of Bryology*).

At the other extreme of the curve of photosynthetic response to light intensity, exposure to high light intensities eventually becomes extremely damaging. Pigment bleaching can occur and the plants may eventually die, a phenomenon all too familiar to anyone who has tried to move a houseplant outdoors into full sunlight. In fact, an excess of light energy that cannot be used for photosynthesis leads to the formation of very strong oxidants that damage the cell structures (Fig. 8.11). This situation, known as oxidative stress (Heber *et al.* 2001), makes it necessary for the plants to regulate light harvest. For example, the leaves of many bryophytes of exposed habitats and, in particular, numerous members of the family Pottiaceae, become strongly crisped or twisted upon drying, thereby mostly exposing the ventral side of the thick costa while protecting the photosynthetic lamina from light. Similarly, in some thalloid liverworts of xeric environments, the margins of the thallus roll up, thereby exposing the ventral side covered by blackish scales and protecting the green dorsal side of the thallus (Fig. 8.12).

Despite the existence of such regulating mechanisms, most plants receive more sunlight than they can actually use for photosynthesis and must face an excess of light energy (Müller *et al.* 2001). Plants therefore need to set up photoprotective mechanisms to dissipate the excess of energy that cannot be used for CO_2 fixation. Non-photochemical quenching (NPQ) of the

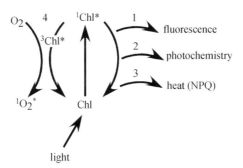

Fig. 8.11. Possible fates of a molecule of chlorophyll (Chl) excited by light energy. When Chl absorbs light, it is excited from its singlet ground state to its singlet excited state, $^1Chl^*$. From there, it has several ways to relax back to the ground state. It can relax by emitting light, seen as fluorescence (1). Its excitation can be used to fuel photosynthetic reactions (2), or it can de-excite by dissipating heat (3). This last reaction reduces the amount of fluorescence and is, therefore, referred to as non-photochemical quenching (NPQ) of Chl fluorescence. Last, $^1Chl^*$ can be transformed into a triplet excited state, which is, in turn, able to produce singlet oxygen $^1O_2^*$, a strong oxidant with very damaging effects for the cell structures (4) (redrawn from Müller *et al.* 2001).

Fig. 8.12. The thalloid liverwort *Targionia hypophylla*, pictured here on volcanic rocks in La Gomera (Canary Islands) in the wet (a) and dry (b) state, thrives in xeric Mediterranean environments. When dry, the margins of the thallus are rolled-up, thereby exposing the ventral side covered by blackish scales and protecting the green dorsal side of the thallus from light (photo D. Aigoin). See plate section for colour version.

chlorophyll molecule through dissipation of the excess energy as heat is one of the most efficient protective mechanisms. Photoprotection through NPQ is regulated by environmental conditions (Beckett *et al.* 2005). When water is freely available and light intensity moderate, NPQ and, therefore, photo-protection, are minimal. Under water stress and high light intensity, the chlorophyll protein complex changes conformation and switches from this very efficient, but unprotected state, towards a less efficient, but photoprotected state characterized by higher NPQ. Photoprotection thus comes with a substantial metabolic cost, so that plants therefore need to continually change the balance of their photosynthetic metabolism between protection and efficiency to track the degree of environmental stress.

8.3 Mineral nutrition

8.3.1 Nutrient uptake and translocation

If desiccation tolerance has several implications regarding the photosynthetic ability of bryophytes, it also impacts on their mineral nutrition. Nutrient uptake and translocation are indeed highly dependent on the plant water status (Sardans & Peñuelas 2008). In an experiment, where nutrients were supplied to plants exposed to different water regimes, Bates (1997), in fact, found that a 24 h dry period imposed every week severely limited the uptake of nutrients.

Because, unlike flowering plants, bryophytes lack roots, it has long been assumed that they largely rely on minerals dissolved in rainwater and

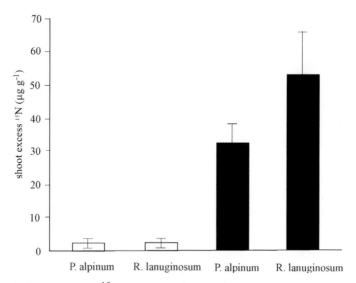

Fig. 8.13. Shoot excess ^{15}N concentration in the two mosses *Polytrichastrum alpinum* and *Racomitrium lanuginosum* exposed to dissolved ^{15}N in soil (white bars) or wet deposition (grey bars). In both the soil and wet deposition treatments, 188 g ^{15}NO$_3$:^{15}NH$_4$ was added in 3.6 ml deionized water. Bars represent means ± standard deviation (reproduced from Ayres *et al.* 2006 with permission of the Royal Society).

aerosols, commonly referred to as 'wet deposition', as well as gases such as ammonia and oxides of nitrogen and dust ('dry deposition'), to sustain their needs. However, in addition to more specialized uptake mechanisms still under scrutiny, such as symbiosis (Sections 1.4.2, 3.1.4, 4.1.4, 5.1.4) and carnivory (see Section 3.1.1), bryophytes are also able to exploit nutrients directly from the substratum. In fact, both ecto- and endohydric species were shown to accumulate ^{15}N when exposed to either soil or wet deposition-derived ^{15}N, clearly demonstrating the potential for upward movement from the soil surface to growing apices (Fig. 8.13). Morphological differences may play an important part in the different uptake strategies. Nutrient acquisition by stiff, erect mosses wherein rhizoids are extremely rare in the wild (Bates & Duckett 2000), such as *Pseudoscleropodium purum* and *Rhytidiadelphus trique-trus*, could mainly rely on precipitation and aerosols for their mineral nutrition. By contrast, mosses with a more plastic shape and a weaker apical dominance, such as the creeping pleurocarp, *Brachythecium rutabulum*, may develop foraging rhizoid-covered runners that permit exploitation of nutrients from the litter (Bates 1997).

8.3.2 Nutrient requirements

A major consequence of the effective mechanism for translocation of nutrients from old to new cells is that the cycling of nutrients is highly efficient in bryophytes. Considerable growth can even occur, temporarily, without further input of inorganic nutrient (Wells & Brown 1996, Bates 1997). Therefore, although bryophytes have, qualitatively, the same nutrient requirements as vascular plants (Bates 2000), they have, quantitatively, extremely low requirements. Their response curve typically exhibits a flat bell-shape, whose optimum corresponds to a solution that is five times less concentrated than the optimum solution used for crop culture (Fig. 8.14). In a series of phosphorus enrichment experiments, Kooijman and Bakker (1993), Steinman (1994) and Martinez-Abaigar *et al.* (2002) failed to find a clear deficiency zone in the physiological response of different species. Although tissue concentrations increased with rising phosphorus supply, no corresponding increase in production occurred (Steinman 1994). Similarly, Wells and Brown (1996) found no evidence of nutrient (potassium, magnesium and calcium) limitation in the moss *Rhytidiadelphus squarrosus*.

At the other extreme of the response curve to nutrients (Fig. 8.14), the negative relationship between fresh weight and nutrient concentration indicates that, beyond a certain concentration, several elements, although required in small concentrations, are toxic when they are present in excess.

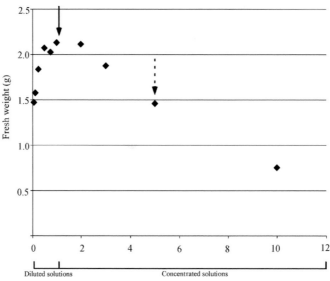

Fig. 8.14. Average fresh weight (g) of clones of the liverwort *Marchantia polymorpha* exposed to increasing nutrient concentrations over 27 days. The x-axis represents the factor of dilution or concentration of the optimal nutrient concentration ($0.012 \, \mathrm{mol \, l^{-1}} \, K^+$, $0.007 \, \mathrm{mol \, l^{-1}} \, Ca^{2+}$, $0.0014 \, \mathrm{mol \, l^{-1}}$ Mg^{2+}, $0.0034 \, \mathrm{mol \, l^{-1}} \, NO_3^-$, $0.004 \, \mathrm{mol \, l^{-1}} \, PO_4^{3-}$ and $0.0008 \, \mathrm{mol \, l^{-1}} \, SO_4^-$) (plain arrow). By comparison, the optimal nutrient concentration used in crop culture is indicated by a dashed arrow (data from Voth 1943).

Mineral elements indeed accumulate within cells and on cell walls. Cell walls possess a net negative charge and therefore bind cations, which can be displaced by other cations from the external medium, particularly if the latter are present at higher concentrations or have higher valences. This ability is called the cation exchange capacity (CEC) and has a major effect on mineral retention.

The CEC regulates the occurrence of species depending on soil conditions, which can be one of the most important factors for bryophyte distributions at a regional scale (Bates 1995). In particular, the failure of calcifuge plants to establish on calcareous soils is usually attributed to the fact that calcium, in excess, causes mineral nutrient deficiencies with regard to other elements such as magnesium. The CEC is three to four times greater in calcicolous bryophytes than calcifuges (Fig. 8.15) (Bates 1982). One interpretation of the absence of calcifuges on calcareous soils is that calcifuges do not possess enough binding sites to address the excess of Ca^{2+} in calcareous environments. However, Bates and Farmer (1990) observed a lack of effect of calcium applications on the growth of *Pleurozium schreberi*, a species that

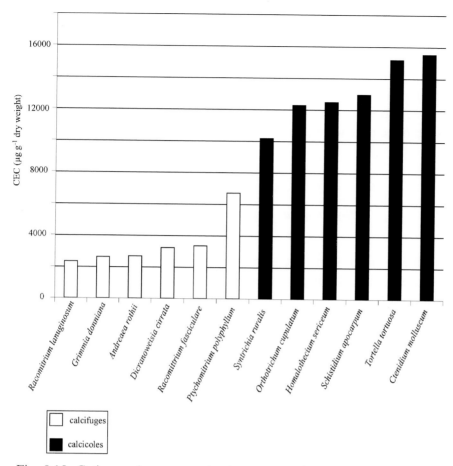

Fig. 8.15. Cation exchange capacity in a range of epilithic calcicole and calcifuge mosses (drawn from Bates 2000).

behaves as a strict calcifuge. Bates and Farmer (1990) therefore suggested that fully developed calcifuge plants may already possess an intracellular capital of essential nutrients and are hence insensitive to changes in external ion supply. Calcium is thus toxic only at the colonizing stage, when the propagules are in intimate contact with the soil and must accumulate a nutrient capital to develop.

A high CEC, which may therefore appear as an advantage for calcicolous plants in calcareous habitats, appears as a limiting factor in acidic environments because of the tendency of calcicoles to concentrate protons and toxic ions (Büscher et al. 1990). This is especially true for aluminium, which is present in the form of highly toxic and mobile Al^{3+} ions in acidic environments. Calcifuges are, by contrast, thought to possess a physiological system, possibly involving specific membrane transfer proteins, which regulates the difference in proton concentration within and outside the cell. Such a mechanism would explain the ability of species, such as the liverwort *Jungermannia vulcanicola*, to grow at a pH as low as 1.9 (Satake *et al.* 1989). In *Sphagnum*-dominated mires, such a mechanism may be responsible for the adsorption of incoming cations and the release of protons, which are added to those already present in the mire water, increasing the acidity of the latter (see Section 2.2).

8.4 Temperature

8.4.1 Physiological temperature optimum and tolerance

8.4.1.1 Temperature optimum

Growth experiments on selected bryophytes of temperate and boreal ecosystems suggest that, although temperature optimum varies among species, from low (12–13°C) in mosses such as *Dichodontium palustre* and *Racomitrium lanuginosum* to relatively high (23–26°C) in, for example, *Bryum capillare* and *Funaria hygrometrica*, the optimum for the majority of species ranges between 15 and 25°C (Furness & Grime 1982b) (Fig. 8.16). Most species exhibit a broad plateau of responses to temperature over the range of

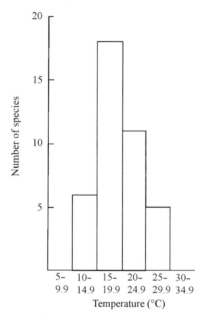

Fig. 8.16. Frequency distribution of the relative growth rate temperature optimum classes in 40 mid-western European bryophyte species (reproduced from Furness & Grime 1982b with permission of Blackwell).

10–25°C. Aquatic bryophytes that develop entirely below the water surface are, by contrast, generally subject to very minor temperature fluctuations and, hence, display a much narrower optimum range than that of terrestrial mosses (Glime & Vitt 1984). In the aquatic moss *Fontinalis antipyretica* for example, net photosynthesis is optimal between 10 and 15°C, beyond which respiration rapidly exceeds photosynthesis (Glime & Vitt 1984).

These temperature optima are lower than those described for temperate vascular plant species. This may reflect the tendency for bryophytes of temperate regions to grow during seasons (spring and autumn) when mois-ture is most readily available and when, coincidentally, temperature is rela-tively low. The moss *Brachythecium rutabulum* displays, for example, a temperature optimum for relative growth rate of 18°C that is substantially lower than that of the tall herb *Urtica dioica* (*c.* 27°C), with which it is commonly associated (Fig. 8.17) (Furness & Grime 1982a). In contrast with the vascular species, *B. rutabulum* still exhibits considerable growth at 5°C. These differences in temperature optimum may account for the co-existence of *B. rutabulum* and *U. dioica*, which exploit different seasonal niches. The

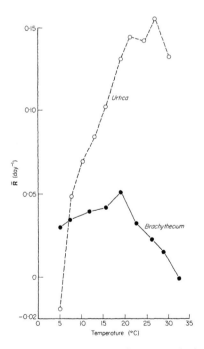

Fig. 8.17. Comparison of the response of mean relative growth rate R to temperature in the tall herb *Urtica dioica* and the moss *Brachythecium rutabulum* grown at 25 W m^{-2} (reproduced from Furness & Grime 1982a with permission of Blackwell).

productivity of *U. dioica* peaks in mid-summer, corresponding to maxima in day length, irradiance and temperature, followed by a rapid decline accelerated by autumn frosts. In contrast, *B. rutabulum* exhibits a complementary phenology, with spring and autumn peaks of productivity. These coincide with the cool, moist conditions at that time of year, when maximum light reaches the moss layer due to the absence of a herbaceous canopy. The comparatively lower temperature optima of bryophytes in comparison with angiosperms also explains certain differences in distribution between the two groups. Along an elevation gradient for example, the maximum of species diversity occurs at a higher altitude in bryophytes than in angiosperms (Section 3.3).

8.4.1.2 *Tolerance to extreme temperatures and relation to water status*

Cold tolerance A feature common among most bryophytes is their ability to grow at low temperatures. More than half of the 40 species investigated by Furness and Grime (1982b) showed a growth reduction of less than 50% at 5°C compared to growth at their optimal temperature. Most species, including tropical ones, seem to be pre-adapted to cold and survive temperatures ranging from -10 to -27°C (Glime 2007a).

As for angiosperms, the tolerance of bryophytes varies seasonally (Rütten & Santarius 1992, 1993). This suggests that mosses develop tolerance in response to changes in environmental conditions. It was experimentally shown that incubation at low, but above freezing temperatures, significantly increases survival rates upon subsequent exposure to negative temperatures. While only 16% of the protonema cells grown at room temperature survived after freezing at -3°C, a cold treatment at 0°C increased the survival rate to 57% after seven days and over 80% after ten days (Fig. 8.18) (Minami *et al.* 2006). The mechanism of freezing tolerance is controlled by ABA (Fig. 8.19). ABA causes an increase in soluble sugars, which play a role in protecting cellular membranes and proteins from freezing injury in much the same way as they do for desiccation tolerance; notably, through the process of vitrification (see Section 8.1.3). ABA also seems to induce the synthesis of proteins for freezing tolerance. Indeed, when the ABA treatment is carried out in the presence of nuclear protein-synthesis inhibitors, freezing tolerance dramatically decreases (Fig. 8.20). When cytoplasmic protein-synthesis inhibitors are used instead, freezing tolerance is not affected, suggesting that nuclear-encoded genes are likely to play a critical role in the development of freezing tolerance. These genes encode for the same LEA proteins that are involved in desiccation tolerance, reinforcing the parallel between the mechanisms involved in desiccation and cold tolerance.

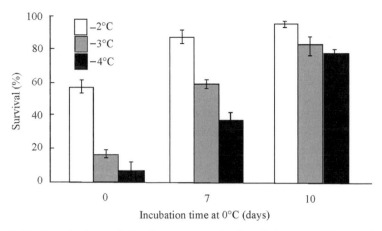

Fig. 8.18. Survival rate (%) of protonemal cells of the moss *Physcomitrella patens* exposed to −2 to −4°C depending on the number of days of incubation at 0°C (reproduced from Minami *et al.* 2006 with permission of CAB).

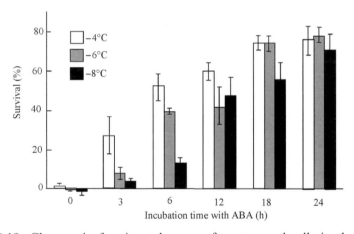

Fig. 8.19. Changes in freezing tolerance of protonemal cells in the moss *Physcomitrella patens* after different periods of incubation on a medium containing 1 μ mol ABA (reproduced from Minami *et al.* 2006 with permission of CAB).

Although bryophytes share with other cryptogams, including ferns, fungi and algae, but not with angiosperms, other ABA-induced genes possibly involved in freezing tolerance, bryophytes and higher plants thus display similarities in their freezing behaviour as well as in the genes required for the development of freezing tolerance. Such evolutionary conservatism with respect to freezing tolerance suggests that these traits might have been necessary for their common ancestor to adapt successfully to non-aquatic

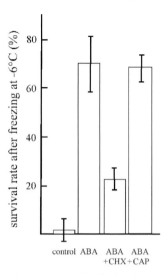

Fig. 8.20. Survival rate of protonemal cells in the moss *Physcomitrella patens* (control), when ABA is added to the growth medium, and when nuclear protein-synthesis inhibitors (CHX) and cytoplasmic protein-synthesis inhibitors (CAP) are added to the growth medium, respectively (reproduced from Minami *et al.* 2006 with permission of CAB).

conditions, and hence have been crucial in the evolution of early land plants (Section 1.4.1).

Heat tolerance At the other extreme of the temperature response curve, all of the temperate and boreal species investigated by Furness and Grime (1982b) died at 35°C and most shoots died eventually at > 30°C. Particularly sensitive species were *Thuidium tamariscinum*, which died after about 20 days at 29°C, and *Chiloscyphus polyanthos*, which died after 5 days at 27°C. Lethality is attributed to damage to membrane systems, including the photosynthetic pigment apparatus. Many bryophytes, though, occur in areas wherein they experience much higher temperatures and must hence have evolved heat tolerance.

In bryophytes, acclimatization of water-saturated shoots by short exposure temperatures above 30°C for a few hours results in a small, albeit significant increase in the thermal stability of the photosynthetic apparatus (Meyer & Santarius 1998) (Fig. 8.21). This suggests that short-term thermal hardening of hydrated tissues may take place in a similar fashion in bryophytes and flowering plants through the production of heat-shock proteins. However, the short-term heat hardening capacity of turgid bryophytes is extremely low and, as opposed to that of angiosperms, may not be ecologically relevant (Meyer & Santarius 1998).

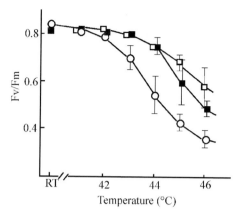

Fig. 8.21. Physiological response, measured as intensity of damage to the photosynthetic apparatus (as assessed by the F_v/F_m ratio), to temperatures above 40°C for 10 min as a function of the length of the acclimation period in the moss *Polytrichastrum formosum*. Shoots were acclimated to 38°C for 2 h (■) and 4 h (□), respectively. Controls (○) were stored at room temperature (reproduced from Meyer & Santarius 1998 with permission of Springer-Verlag).

In bryophytes, heat tolerance strongly increases in the dry state. Under natural conditions, in fact, mosses lose water rapidly when temperature increases, in parallel with a drop in relative humidity, which is accompanied by a prompt rise in thermotolerance. Whereas the lethal thermal limit for metabolically active (i.e. hydrated) gametophytes is 51°C, this limit increases to 110°C when gametophytes are desiccated (Meyer & Santarius 1998).

Differences in response to heat exposure also vary between gametophytes and sporophytes. While all maternal gametophytes of the moss *Microbryum starckeanum* survive exposures up to 75°C for 1–3 h, no embryonic sporophyte remains viable at that temperature (McLetchie & Stark 2006). This suggests that either the inherent thermotolerance is truly lower in the sporophyte than in the gametophyte, which is consistent with the observation that sporophyte production is restricted to the coolest, wettest months, or that gametophytic thermal stress response controls sporophyte viability.

8.4.2 Consequences for species distributions and phenology

Temperature appears as a factor of prime importance in bryophyte physiology. In fact, temperature seems to regulate the succession of a suite of complementary mechanisms regarding growth and reproduction. For example, the phenology of the aquatic moss *Fontinalis* follows seasonal temperature variation (Glime 1982). In the spring, growth of *Fontinalis* increases as temperature rises to 10°C. During that period, branching is initiated and rapid branch growth occurs, enhancing photosynthesis in cold water, rich in dissolved CO_2. In the summer, growth and branching continue until the temperature reaches 15°C. As it approaches 20°C, growth nearly stops. By contrast, rhizoid production increases, causing firmer attachment, preparing the shoot for abrasion caused by autumn rains and spring snow-melt. The highly resistant stem is thereby maintained so that growth can resume when temperatures drop in the autumn. Archegonia and antheridia develop by early autumn. Capsules are produced on older branches that lack leaves and are easily detached by abrasion. During the winter and early spring, the run-off following snowmelt causes extensive abrasion, favouring dispersal. Each season thus appears uniquely necessary in the *Fontinalis* cycle, leading Glime (1982) to suggest that the moss cannot extend its range into a climate with no temperature seasonality. This would explain, for example, the rarity of aquatic bryophytes in the tropics, even in habitats such as low-elevation rapids and waterfalls, which would be massively colonized at higher latitudes (Glime & Vitt 1984).

At a finer scale, however, the relationships between optimal growth responses to temperature in experimental conditions and species distributions are, most often, not evident. For example, in their analysis of 40 temperate and boreal bryophytes, Furness & Grime (1982b) failed to find a relationship between growth response to temperature and altitude of origin. A possible explanation for this counterintuitive observation is that the microclimate experienced by bryophytes may be very different from the gross climatic characteristics of a region. Bryophytes of geothermal habitats, with highly contrasting air and ground temperatures (Convey & Lewis-Smith 2006), or hyper-oceanic bryophytes restricted to micro-refugia within continental areas, are good examples of the mismatch between species occurrences and global climate at the regional scale.

The success of bryophytes in the field may also be related to effects, which are relatively independent of the temperature optimum, such as the capacity to grow at low temperatures or, conversely, to survive episodes of extreme heat. Many species are capable of net positive growth at temperatures as low as 1°C, allowing them to maintain vegetative growth almost throughout the year. This enables them to colonize high altitudes or polar regions. Yet, thermotolerance might still be limiting in alpine areas, wherein a cushion of *Grimmia* may experience temperatures up to 60°C during the day and well below freezing at night.

It is also possible that bryophytes are physiologically plastic, i.e. are capable of modifying their physiology in response to environmental variation,

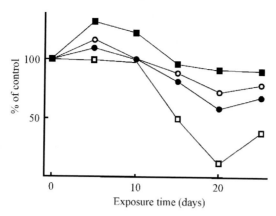

Fig. 8.22. Variation in respiration rates of the aquatic moss, *Fontinalis antipyretica*, exposed to increasing temperatures (■ 19.8°C, ○ 24.9°C, ● 28.4°C, □ 33.5°C) during a 25 day period, as expressed in per cent of control values obtained at 16°C (redrawn from Carballeira *et al.* 1998).

which might allow them to adapt to hostile conditions. Even the aquatic moss *Fontinalis antipyretica*, which displays a fairly narrow range of optimum temperature, is capable of acclimatization to moderately high temperatures. Although this moss normally exhibits an optimum around 15°C, a period at 20–25°C eventually results in a shift in the compensation point, i.e. in a reduction of the respiration rate at the new temperature, apparent after about 10 days (Fig. 8.22) (Carballeira *et al.* 1998).

9

Bryophytes in a changing world

The natural corollary of the narrow relationship between bryophytes and their environment is that a great deal of variation in community composition and physiological behaviour is to be expected whenever changes in environmental conditions occur. In the current context, global change brought about by human activities has increasingly been an area of concern. Bryophytes, perhaps more than many other organisms, are highly sensitive. Their response to climate change is obvious from the reconstruction of paleovegetation based on the analysis of macrofossils (see Section 7.1), so that a great deal of changes in species distribution and community composition is to be expected in the current context of global warming (Box 7.1). The response of bryophytes to an increasingly polluted world is no less dramatic. This is because bryophytes, being poikilohydric, mostly lack a protective, highly hydrophobic cuticle and are, therefore, directly exposed to pollutants. As a result, pollutants may bind on cell walls, eventually pass through the cell membrane and affect metabolism.

In this chapter, we review how pollution affects bryophyte ecophysiology and community composition. We focus on airborne pollution, which arguably constitutes the main threat to ecosystems (Lee 1998). We then assess how understanding the response of bryophytes to these changes can be used as one indicator to monitor the quality of the environment. Finally, we examine the mechanisms by which bryophytes are able to adapt and survive in a changing environment.

9.1 Impact of pollution on bryophytes

Nitrogen and sulphur are essential elements for all organisms. Although the ecological importance of their natural biogeochemical cycles has long been recognized, little emphasis has been placed until recently on the ecological consequences of the perturbation of these cycles by human activities. A realization of the scale of these perturbations and of their potential ecological effects has just begun to focus attention on them. In particular, the spectacular forest decline in industrial countries during the 1980s and the 1990s focused the attention on the dramatic impact of acidic mists and rains largely caused by SO_2 pollution. SO_2 concentrations sharply increased as a result of developing smelting activities in the immediate pre-industrial era. In the second half of the twentieth century, changes in legislation, fuel usage and combustion technology resulted in the increasing importance of high level emissions from power stations. This resulted in increased dispersion of SO_2 and much lower concentrations in formerly polluted districts. These changes in fuel usage and combustion technology are one of the factors resulting in increasing atmospheric nitrogen deposition, but may also have exacerbated the dispersion of acidic pollutants such as NO_3^-. The successive waves of nitrogen and sulphur pollution cannot, therefore, be looked upon as being entirely independent of one another. Long-term changes induced by sulphur pollution may still be occurring in ecosystems as their responses to nitrogen are initiated and these changes interact with one another (Lee 1998). It is the impact of these ecological changes on bryophytes that are the focus of the next section.

9.1.1 Acidification

Long-term exposure to acidic precipitation has devastating effects on the photosynthesis, growth and survival of bryophytes. Forest bryophytes experimentally sprayed with acid water started to show a linear decline in dry weight from pH < 4 (Fig. 9.1). Significant reductions of up to 75% in both frond height and dry weight were recorded at pH 3.5 and lower (Hutchinson & Scott 1988). The decline in growth of bryophytes exposed to experimental acid rains is highly correlated with a significant reduction in chlorophyll content, up to 35% at pH 2.5 and 3.0 (Fig. 9.2) (Bakken 1993). The high H^+ concentrations may induce the hydrolysis of chlorophyll *a* by displacing the Mg^{2+} ions from the chlorophyll molecule and converting it to phaeophytin. Acid rain may also reduce the available magnesium for chlorophyll synthesis by washing off dry depositions, including Mg^{2+}, from the moss surfaces.

While definite positive trends could be observed during the past two decades regarding the emission of SO_2, accompanied by a remarkable recolonization of formerly polluted areas by sensitive species (Bates *et al.* 1997), the consequences of past acid rains can still be observed today in a wide range of bryophyte communities. The most dramatic vegetation changes attributed to

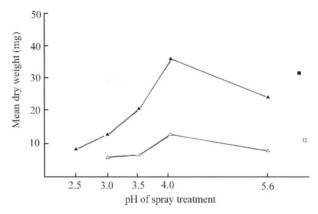

Fig. 9.1. Mean dry weight (mg) of fronds of the moss *Pleurozium schreberi* plotted against pH of a simulated rain treatment, delivered twice-monthly throughout the growing season over 5 yrs. ▲ mean total dry weight; Δ mean dry weight of the green, photosynthetically active portion of the frond; ■ total frond dry weight, control; □ living frond dry weight, control (repro-duced from Hutchinson & Scott 1988 with permission of *Canadian Journal of Botany*).

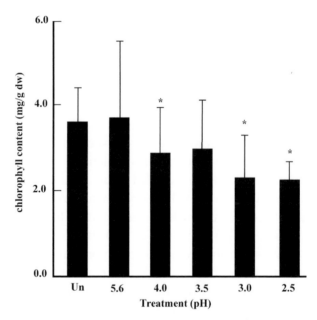

Fig. 9.2. Mean chlorophyll content (and SD bars) of the moss *Hylocomium splendens* after two seasons of exposure to simulated acid rain (25 mm twice a month from May to October) at five pH levels and unsprayed (Un) as controls. Asterisk indicates significantly different mean content from control (reproduced from Bakken 1993 with permission of *Lindbergia*).

acidification have taken place in habitats that, due to their low buffering capacity, experienced a rapid and substantial decrease in pH. In epiphytes, for example, communities of rich bark composed of *Orthotrichum* spp. and *Cryphaea heteromalla* shifted towards acidophilous communities dominated by Dicranaceae (e.g. *Dicranum tauricum* and *D. montanum*), which have drastically increased in abundance since the middle of the twentieth century (Bates *et al.* 1997). Similarly, in streams, acidification caused a shift in community composition (Stephenson *et al.* 1995, Thiébaut *et al.* 1998). A set of species characteristic of neutral waters and sensitive to acidity from the very first stages of their development, such as *Platyhypnidium riparioides* and *Chiloscyphus polyanthos* (Fig. 9.3), progressively disappeared. In contrast, acidophilous species, such as *Scapania undulata*, are physiologically equipped to face the proton load (Section 8.3.2), hence exhibit greater tolerance to acidic environments (Fig. 9.3), and have therefore spread. Perhaps the most spectacular effects of acidification through SO_2 pollution, however, have taken place in ombrotrophic mires. These ecosystems are especially susceptible to atmospheric pollution because they receive the majority of their mineral supply from the atmosphere (Hogg *et al.* 1995). Although acidification is part of a natural, autogenic evolutionary process in bogs (Section 2.2), succession from mineral-rich fens towards more acid stages dominated by *Sphagnum* has been occurring on a rapid scale in areas highly exposed to acid deposition. In the most polluted areas, acid rains have changed the course and velocity of succession towards communities, in which *Sphagnum* itself is absent. For example, the virtual absence of *Sphagnum* from the southern Pennines, an area of more than 50 000 ha, is a remarkable feature of British vegetation because it remains the only case in which experimental and observational evidence can be combined to demonstrate the ecological importance of SO_2 pollution at a large scale (Lee 1998). In the Pennines, the disappearance of *Sphagnum* in peat stratigraphic profiles indeed correlates with the appearance of soot from industrial origin in the deposits (Tallis 1964). Fumigation experiments in closed chambers further demonstrate that *Sphagnum* growth is significantly reduced when SO_2 concentrations reach $131 \, \mu g \, m^{-3}$, a concentration within the range observed during the periods of highest levels of pollution in the twentieth century (Ferguson *et al.* 1978). Finally, field experiments involving the spraying of dilute concentrations of bisulphite ions onto a *Sphagnum* bog surface resulted in the removal of the *Sphagnum* cover within a year (Ferguson & Lee 1980). Collectively, all these experiments demonstrate that the disappearance of *Sphagnum* from the Pennines can in fact be attributed to the levels of SO_2 pollution that the area experienced during the first half of the twentieth century.

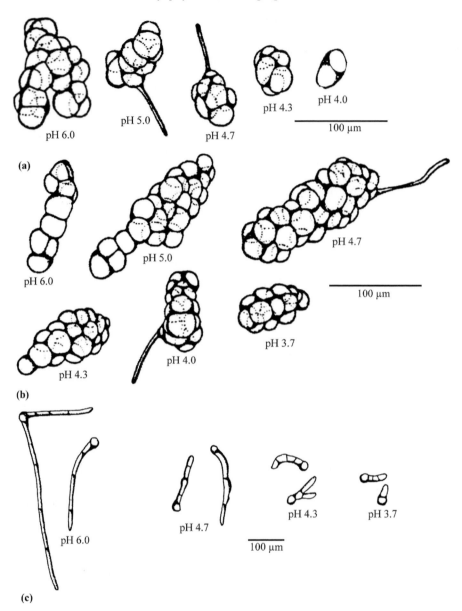

Fig. 9.3. Contrasting effects of pH on the development of sporelings in the leafy liverworts *Chiloscyphus polyanthos* (a) and *Scapania undulata* (b) and of the protonema of the moss *Platyhypnidium riparioides* (c) (redrawn from Tremp & Kohler 1993).

9.1.2 Eutrophication

By contrast with SO_2, several other compounds, particularly nitrogen oxides, have been drastically increasing during the past several decades as a result of farming, fertilizer production and fossil fuel burning, resulting in an alarming eutrophication of the environment. This trend is well illustrated by the two- to three-fold increase in nitrogen concentrations during the twentieth century that has been reported in moss tissues from herbarium collections (Fig. 9.4) (Woolgrove & Woodin 1996). Although several countries have expressed the intention to reduce nitrogen emissions (see UN/ECE-protocol from Gothenburg at www.unece.org/env/lrtap), this trend is likely to continue as a result of fossil fuel consumption and fertilizer use and to lead to a projected 60% increase by 2020, with dramatic consequences for ecosystems (Lee 1998).

High nutrient concentrations can cause toxic effects directly, the mechanisms of which are still under scrutiny (Koranda *et al.* 2007). Lethal ammonium concentrations range from $7.5\,\mathrm{mg\,l^{-1}}$ to $500\,\mathrm{mg\,l^{-1}}$ in a variety of aquatic mosses (Vanderpoorten 2003). Such tolerances lie within the range of pollution levels that existed in large European rivers during the 1960s and 1970s that peaked at concentrations reaching $20\,\mathrm{mg\,l^{-1}}$ and beyond. Since then, water quality in large rivers has globally drastically improved, so that the aquatic bryoflora is, exactly as epiphytes, in the process of re-invading

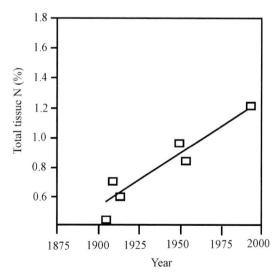

Fig. 9.4. Variation of tissue N concentrations (% dry weight) in herbarium material of the moss *Kiaeria starkei* in a Scottish snowbed between 1875 and 2000 (reproduced from Woolgrove & Woodin 1996 with permission of Elsevier).

rivers from where it almost completely vanished during the 1970s (Vander-poorten 1999).

The most substantial effects of nitrogen are, however, indirect. Nitrogen is the limiting nutrient for plant growth in many ecosystems and is therefore an important factor determining the vegetation composition. Many plant species are adapted to nutrient-poor conditions and can only compete successfully on soils with low nitrogen levels (see Bobbink *et al.* 1998 for review). Unlike most angiosperms, bryophytes are not limited by nutrient availability in many instances, perhaps as a result of their highly efficient cycling of nutrients (Section 8.3.2). Although bryophytes from certain habitats, especially forests, may exhibit slow responses to increased nitrogen deposition (Zechmeister *et al.* 2007b), an increase in nutrients generally permits a few fast growing, tall species to profit the most from the additional nutrients. These plants expand rapidly and overshadow species that have lower potential growth rates or that, for other reasons, are unable to profit from the extra nutrients available (Berendse 1999, Pearce & van der Wal 2002). In agricultural meadows, for instance, the substantial decrease in bryophyte diversity that has been observed during the twentieth century is directly correlated with nitrogen loads (Zechmeister *et al.* 2003a) (Fig. 9.5). Another good example is given by the increasing dominance of *Sphagnum fallax* to the detriment of *S. papillosum* and *S. magellanicum* over the last century in The Netherlands (Limpens *et al.* 2003). Dominance of *S. fallax* over the other species occurs only in bogs with a sufficiently high nutrient availability (Fig. 9.6). Under such conditions, the length increase in *S. fallax* is significantly higher than in the two other species (Fig. 9.6), resulting in the elevation of the capitula above the moss carpet (Fig. 9.7a). Then, as support from the surrounding sphagna diminishes, the top-heavy and often branching *S. fallax* individuals collapse over the surrounding *S. magellanicum* or *S. papillosum*, ultimately shading them out (Fig. 9.7b). Once dominant, the relatively high degrad-ability of *S. fallax* tissue is likely to help maintain the high nutrient availability in the environment, thus securing an extended period of supremacy (Limpens *et al.* 2003).

Negative impacts of nitrogen deposition are further enhanced in the con-text of global climate change. Manipulative experiments in open ecosystems such as bogs and grasslands indicate that fertilizer addition and increased temperatures create rapid responses (within 1–3 years) in graminoids that eventually overshadow the bryophyte layer (Molau & Alatalo 1998, Berendse *et al.* 2001, Bergamini & Pauli 2001). In Arctic regions, in particular, warming is expected to increase the activity of soil micro-organisms and, hence, rates of mineralization of organic mater. Because plants may benefit from

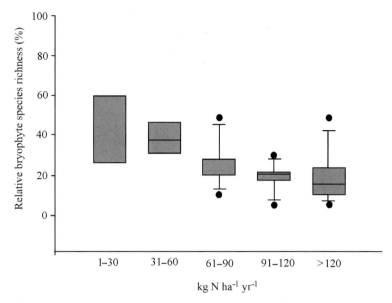

Fig. 9.5. Impact of nitrogen inputs on bryophyte diversity in agricultural meadows. Bryophyte diversity is expressed as a per cent of the total species diversity expected in untreated meadows (reproduced from Zechmeister *et al.* 2003a with permission of Elsevier).

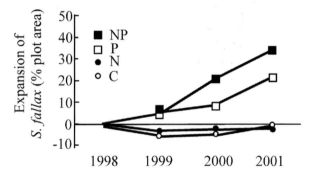

Fig. 9.6. Effects of three years of N and P fertilization on the expansion of *Sphagnum fallax* in *S. magellanicum* and *S. papillosum* vegetation, expressed as mean change (±standard deviation) in covered plot area. C = control, N = 40 kg N ha^{-1} yr^{-1}, P = 3 kg P ha^{-1} yr^{-1} (reproduced from Limpens *et al.* 2003 with permission of *Journal of Bryology*).

eutrophication only if temperature increases, the interaction between these two factors is likely to have a greater impact on bryophytes than any of these factors considered alone (Figs. 9.8, 9.9).

Competition for light is, however, not necessarily the only factor that reduces bryophyte growth under eutrophic conditions. In The Netherlands,

Fig. 9.7. Expansion of *Sphagnum fallax* induced by addition of P (see Fig. 9.6). (a) *S. fallax* capitula (1) thrusting out of *S. magellanicum* (2). (b) Continued expansion of *S. fallax* through frequent branching (1), leading to overtopping and subsequent shading out of *S. papillosum* (reproduced from Limpens *et al.* 2003 with permission of *Journal of Bryology*). See plate section for colour version.

for instance, characteristic species of rich fens, such as *Scorpidium scorpioides*, have been replaced by the weedy *Calliergonella cuspidata* in areas characterized by significantly higher concentrations of phosphorus and greater biomass of vascular plants (Kooijman 1993). Yet, *S. scorpioides* is more stimulated by increased nutrient availability and exhibits a lower compensation and saturation point than *C. cuspidata* (Kooijman & Bakker 1993).

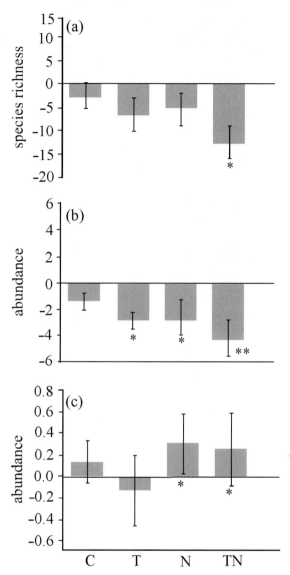

Fig. 9.8. Changes in mean (with 95% confidence interval) bryophyte species richness (per 100 cm^2) (a) and abundance (b) (as assessed by the frequency in each of 18 10 × 10 cm plots) and graminoid abundance (c) from 2000 to 2003 in a Norwegian alpine *Dryas octopetala* heath depending on temperature and/or nutrient supply. Temperature was increased by an average of 9.0 ± 0.1°C to 10 ± 0.1°C. Fertilization consisted of a supply of 10 g N, 2 g P, 8 g K per m^2 per growing season. C = control, T = temperature increase, N = nutrient supply, TN = temperature increase and nutrient supply. Stars indicate significant average differences (reproduced from Klanderud & Totland 2005 with permission of *Ecology*).

Fig. 9.9. Four years of experimental warming and nutrient addition (see Fig. 9.8) altered the high-diversity *Dryas octopetala* heath (above) to a low-diversity community dominated by graminoids (below) (reproduced from Klanderud & Totland 2005 with permission of *Ecology*). See plate section for colour version.

The dominance of the latter seems induced rather by the increasing amount of litter that results from the increase in biomass in the herb layer. *Scorpidium scorpioides* tolerates neither desiccation nor a cover of litter. By contrast, the more desiccation-tolerant *C. cuspidata* exhibits a 'foraging strategy'. Aided by a high degree of morphological plasticity in the rate and direction of shoot proliferation, *C. cuspidata* seems able to take advantage of the climbing frame offered by the litter and to benefit from the increased availability of CO_2 in the above-litter drier environment (Kooijman & Bakker 1993).

9.2 Application to bioindication

Because bryophytes react in predictable and measurable ways to changes in their environment, they have been used as bioindicators. Bioindicators are organisms, or communities of organisms, which supply information on the quality of the environment through some aspect of their biology (Markert *et al.* 2003). The idea of using organisms to register and evaluate certain characteristics of the environment is based on the congruence between the prominence of certain environmental factors and species requirements and can be traced back to the sixteenth century (Markert *et al.* 2003). Certain forms of plant cover were already known to indicate the presence of ore in the ground and the composition of the vegetation was used to assess soil fertility.

With the beginning of the industrial era, bioindication strongly developed as a response to an increasing political demand to monitor the evolution of pollution levels and take the appropriate measures. Bioindicators exhibit a series of complementary features that traditional chemical analyses do not offer. In particular, pollution loads experience dramatic fluctuations in space and time, but also in the nature of the pollutants involved, making it extremely difficult to know which pollutants to measure, where and when. In contrast, bioindicators may react to a wide range of pollutants and, because they are permanently exposed to the environmental pollution, integrate the pollution events over time.

Two main classes of biomonitoring techniques, direct and indirect, can be distinguished. Direct biomonitoring techniques use organisms as *in situ* biological assays of pollution levels. This is how bryophytes are most often used in biomonitoring, with pollutant concentrations being directly measured in moss tissues. Sometimes, indicators of plant stress, such as pigment ratios (Lopez *et al.* 1997) or the concentration of enzymes involved in the detoxification process of pollutants (Roy *et al.* 1995, Schrenk *et al.* 1998), are also used. Indirect biomonitoring, by contrast, involves detecting how variation in community attributes, such as species composition, abundance and diversity, may be explained by changes in the environment.

9.2.1 Indirect biomonitoring

Indirect biomonitoring relies on differences in the sensitivity to pollutants among species to characterize the quality of the environment. One of the most extensively documented applications in bryophytes is the estimation of air quality from the composition of epiphytic species assemblages. This approach was initially introduced by LeBlanc and De Sloover (1970) through the 'index of atmospheric purity' (IAP). IAP is directly related to the bryophyte species richness and is the product of the ecological index of each species and its cover at the study site, summed over all species at the study site. In the absence of detailed information on the response of individual species to pollution, LeBlanc and De Sloover (1970) proposed that the ecological index of each species (i.e. its sensitivity to pollutants) can be established by adding together the number of species in the community, then taking the average of

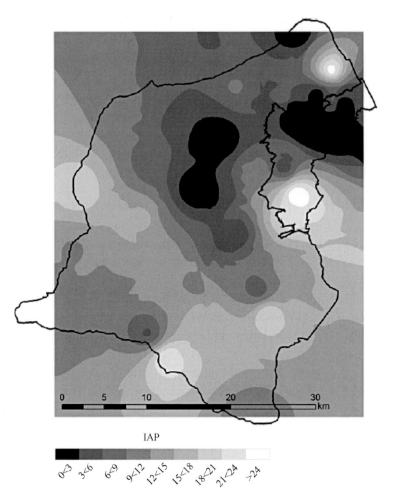

Fig. 9.10. Isoclines of global air quality in the vicinity of the city of Vienna (Austria) derived from an index of air purity (IAP) based on the occurrence of indicator epiphytic species (reproduced from Krommer *et al.* 2007 with permission of Elsevier). Isoclines were deduced from observations made over 30 sites. The average IAP value of 10.6 corresponds to $16.8\,\mu g\,NO_2\,m^{-3}$, $6.1\,\mu g\,SO_2\,m^{-3}$ and $24.2\,\mu g$ dust particles m^{-3}.

the sums for all sites where the species is present. Ideally, a specific ecological index should be calculated for each area, since the same species does not necessarily exhibit the same sensitivity from one area to another (Zechmeister & Hohenwallner 2006). The IAP of all investigated sites can then be mapped to produce a graphic representation of the pollution loads, which can be used to follow changes in the spatial extent of pollution through time (Fig. 9.10) (Zechmeister & Hohenwallner 2006, Krommer *et al.* 2007).

9.2.2 Direct biomonitoring

Bryophytes are ideal candidates for direct biomonitoring because of their extreme capacity to accumulate pollutants such as heavy metals, radioisotopes and a wide range of chemical pollutants such as dioxins (Zechmeister *et al.* 2003b, Carballeira *et al.* 2006). The bioconcentration factor, defined as the ratio between the concentration in the moss ($\mu g\ g^{-1}$ dry weight) and concentration in the environment, is indeed exceptionally high in bryophytes. For example, bioaccumulation factors of 8 400 000 for iron and 11 250 000 for manganese have been reported in mosses from areas highly exposed to heavy metal pollution (Ah-Peng & Rausch de Traubenberg 2004).

The reasons for this extreme intrinsic capacity of bryophytes to over-accumulate are many. Bryophytes lack the complex regulatory mechanisms of vascular plants. In bryophytes, gas exchange is not regulated by stomata and nutrients are not pumped up by the root system. Rather, bryophytes readily absorb pollutants from their immediate environment through their surfaces. The leaves, which mostly lack a protective cuticle, are most often one cell thick and therefore offer a large surface of absorption.

Pollutants accumulate both within and among cells. They enter and leave the tissues more readily if they are outside the cells than if they are inside. As a result, the extracellular fraction reflects current pollution levels in the environment, whereas the intracellular fraction is more constant and testifies to the average pollution load that is present in the environment (Mouvet & Clavieri 1999). The intracellular concentration of pollutants thus retains the 'memory' of past pollution events and may testify to former pollution peaks that are no longer detectable in the environment (Mouvet *et al.* 1993). This property is especially useful for monitoring effluent from nuclear power plants, for which instantaneous pollution levels remain mostly below detection levels (Beaugelin-Seiller *et al.* 1994).

Moss analyses make it possible to determine patterns of pollution loads at large geographical scales and to identify the most heavily polluted areas. Such analyses also make understanding of long-range pollution much faster and cheaper than using continuous chemical analyses of precipitation. In Europe, for example, the Heavy Metal Deposition Programme involves measurements of concentrations of ten heavy metals in naturally growing mosses at five-year intervals (Harmens *et al.* 2004). Twenty-eight countries were involved in the 2000–2001 survey with a total of about 7000 sites investigated. Heavy metal concentrations in mosses increase eastwards, which can be related to industrial emissions (Fig. 9.11). The elevated concentrations of heavy metals in areas without current emission sources, such as lead in southern Scandinavia,

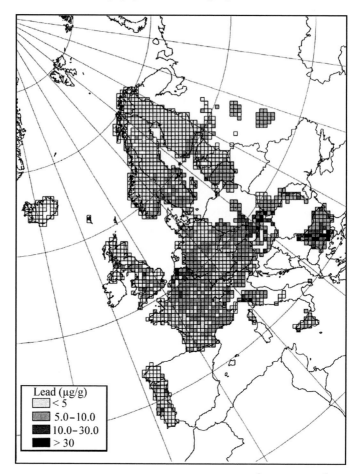

Fig. 9.11. Mean concentration of Pb in moss ($\mu g\,g^{-1}$) per 50 km^2 in Europe, as assessed during the 2000–2001 Heavy Metal Deposition Programme (reproduced from Harmens *et al.* 2004 with permission of Kluwer Academic Publishers).

suggest long-range transboundary transport of pollutants from emission sources elsewhere in Europe.

An alternative strategy to the measurement of heavy metals directly in mosses found *in situ* is the use of moss bags. Samples of mosses collected from clean areas are placed in nylon nets and exposed to a polluted area to enable concentrations in pollutants to be measured afterwards. The technique can be applied in environments, where mosses do not naturally occur. As an extreme example, mosses have been used as indicators of polycyclic aromatic hydrocarbon pollution resulting from the incomplete combustion of organic material, and fossil fuels in particular, in road tunnels (Zechmeister *et al.* 2006).

9.3 Adaptation to a changing environment

Like all living organisms, bryophyte species tend to differ in ecological requirements (Chapter 7) and are, hence, reliable indicators of habitat conditions and pollution levels. We would therefore expect that at least some species have adapted to specific environments. A high degree of niche conservatism is found in a least some major bryophyte lineages (Section 4.3), suggesting that their ecological range is phylogenetically constrained.

In plant populations, the evolution of metal tolerance has become a model system for studying microevolutionary processes (Shaw 1994). Heavy metals such as copper, zinc, lead and cadmium are extremely toxic to organisms (both plants and animals) when present in greater than trace concentrations, because they inhibit basic physiological processes, including photosynthesis and respiration (Fig. 9.12) (Brown & Whitehead 1986, Brown & Wells 1990). For some metals, mercury in particular, inhibition occurs even at extremely low concentrations ($5 \mu M$) and photosynthesis was found to be 80% inhibited after 1 h of exposure to $25 \mu M$ Hg (Brown & Whitehead 1986).

Because heavy metals are so toxic to virtually all plants and animals, their presence in the environment results in strong natural selection favouring tolerant variants. Research on a variety of angiosperms has shown that populations growing in contaminated environments generally represent genetically specialized tolerant races (Shaw 1994). Certain bryophytes are also known to occur in habitats with high concentrations of heavy metals. For example, the so-called 'copper moss' *Scopelophila cataractae* is restricted to soils highly contaminated by heavy metals (Shaw 1987). Growth of *S. cataractae* is promoted by copper and the species exhibits a reduced vigour on uncontaminated

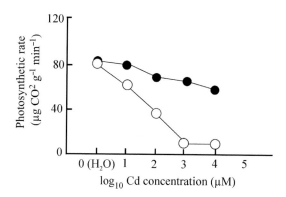

Fig. 9.12. Photosynthesis rate in the moss *Rhytidiadelphus squarrosus* determined immediately following (●) or 24 h after (○) a 30 min incubation along a gradient of exposure to Cd (reproduced from Brown & Wells 1990 with permission of Oxford University Press).

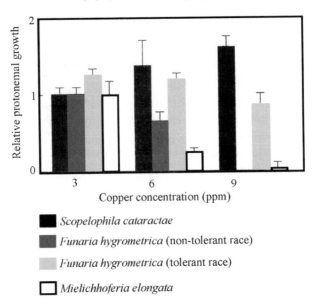

Fig. 9.13. Mean growth (with standard deviation bars), expressed as a percentage of growth on control medium, of *Scopelophila cataractae*, *Mielichhoferia elongata* and two races (copper tolerant and non-tolerant, respectively) of *Funaria hygrometrica* on media supplemented with 3, 6 or 9 ppm copper (reproduced from Shaw & Owens 1995 with permission of *Fragmenta Floristica et Geobotanica*).

soils (Fig. 9.13). Metal tolerance can also be found in species that do not systematically occur on highly contaminated soils. In the liverwort *Marchantia polymorpha* and in the moss *Funaria hygrometrica*, for example, populations vary in their degree of tolerance to copper (Fig. 9.13) and these differences are directly related to the levels of soil contamination (see Shaw 1988 for review). These observations are in accord with those from numerous studies of flowering plants, in which high levels of tolerance to metals are almost always restricted to plants growing in contaminated environments. By contrast with *S. cataractae* and other metal-tolerant angiosperms and fungi, however, tolerant populations of *F. hygrometrica* do not seem to be counter-selected on non-contaminated soils (Fig. 9.13). The observation that the degree of tolerance among populations is correlated with substrate metal concentration suggests, however, that there is some as yet undetected disadvantage for tolerant plants on normal soils (Shaw 1988).

In many other instances, by contrast, surprisingly little evidence of ecotypic differentiation was found in bryophytes (Shaw 2000). In *Bryum argenteum*, for example, plants from clean and heavily polluted environments exhibit indistinguishable growth responses to media supplemented with heavy metals. The degree of tolerance to copper and zinc is intermediate between the levels of tolerance exhibited by tolerant and non-tolerant races of *Funaria hygrometrica*. It may be that this inherent, relatively high level of tolerance makes the evolution of specialized races unnecessary (Shaw *et al.* 1989, Shaw & Albright 1990). Although bryophytes are not genetically depauperate and, in fact, display amounts of genetic diversity comparable with angiosperms (Hedderson & Nowell 2006), this genetic variation does not appear to be adaptive to specific environments (Shaw 1992). Altogether, these observations suggest that, by contrast with flowering plants, physiological acclimation is much more important for bryophytes than genetic specialization.

10

Conservation biology

The major environmental changes that bryophytes are facing (Chapter 10) have increasingly brought about concerns for their conservation. Conservation biology is a fairly new, multidisciplinary science that has developed to deal with the crisis confronting biological diversity (Primack 1993). As a crisis discipline, conservation biology appeared in response to a rising political demand to face the dramatic loss of biodiversity and the need to take steps to anticipate, prevent and reverse the trend (Heywood & Iriondo 2003). Subsequent ratification of the 'Convention on Biological Diversity' at the United Nations conference held in Rio in 1992 by most of the world's governments has placed the subject of biodiversity firmly on the political agenda.

The development of the concept of conservation biology is closely interrelated with that of biodiversity. Biodiversity is defined in the Convention on Biological Diversity as 'the variability among living organisms, from all sources . . . and the ecological complexes of which they are part'.

In this broad conservation picture, bryophytes were initially completely overlooked. Cryptogams in general were, for example, the focus of only about 4% of published papers between 2000 and 2005 in leading conservation journals (Hylander & Jonsson 2007). The situation has, however, been changing recently and there has been an increasing awareness of the necessity to include cryptogams in general, and bryophytes in particular, in conservation programmes (Hylander & Jonsson 2007).

The reasons for a late but growing interest in bryophyte conservation are many. Although bryophytes are rarely the most conspicuous elements in the landscape, they play important ecological roles in terms of water balance, erosion control and nitrogen budget, as well as simply by providing habitats for other organisms (Chapter 2). Furthermore, bryophytes locally exhibit richness levels that are comparable to, or even higher than, those of angiosperms. In boreal forests, for example, bryophyte diversity often exceeds that of vascular plants at a scale of 0.1 ha (Berglund & Jonsson 2001). In the wet sclerophyllous forest of Tasmania, the ratio of the number of bryophyte to vascular plant species is about 5:1 (Pharo & Blanks 2000). Lastly, and perhaps most importantly, diversity patterns in bryophytes do not necessarily follow the patterns present in other, better studied, taxa at both a global (see Section 6.1) and local scale (Sérgio *et al.* 2000, Pharo *et al.* 2005), so that an enlarged concept of biodiversity has become increasingly necessary.

In this final chapter, we review the tools that are available for assessing threat levels in bryophytes. We then provide an overview of global threat levels and conservation needs and review the mechanisms by which bryophytes are, at least locally, severely threatened. Finally, we discuss appropriate conservation and restoration strategies for preserving, managing and, if necessary, reintroducing bryophyte diversity.

10.1 Levels of threats and the need for conservation

10.1.1 What to conserve? A hierarchical system of threat categories applied to bryophytes

Conservation, 'the philosophy of managing the environment in such a way that does not despoil, exhaust, or extinguish it' (Jordan 1995), is by definition concerned with the threat of extinction of species, communities or ecosystems due to human activities. To date, the number of species believed to be under some degree of threat makes necessary the use of a system of classification that helps categorize species according to the risk of extinction they face (Heywood & Iriondo 2003).

The likelihood of extinction of a species must be assessed against certain criteria. This is the purpose of a red list. In a red list, the species found in a given area are ranked according to their threat levels, which can be assessed thanks to the guidelines of the International Union for the Conservation of Nature and Natural Resources (IUCN; IUCN 2001). Five quantitative criteria are estimated to determine whether a taxon is threatened or not (Table 10.1) and, if so, which of the seven threat categories it belongs to (Box 10.1).

10.1.2 Level of threat in the bryophyte floras

The vast majority of bryophyte species are sparsely distributed, within a given area or globally (Cleavitt 2005). Species frequency distributions are typically highly skewed, with rare and potentially threatened species representing the bulk of the flora (Fig. 10.1) (Longton & Hedderson 2000). Figure 10.2 shows

Table 10.1. *IUCN criteria of species threat categories (IUCN 2001)*

Criterion	Threshold
Declining population	30–90% population decline during a time period of 10 years or 3 generations, whichever is longer
Rarity and decline	$EOO^{(1)} < 20\,000\,km^2$ or $AOO^{(2)} < 2000\,km^2$ and severe fragmentation, continuing decline or extreme fluctuations
Small population size and fragmentation, decline or fluctuations	Population size < 250 reproductive individuals and continuing decline of $> 10\%$ in 10 years or 3 generations
Very small population size or very restricted distribution	Number of individuals < 1000 or $AOO < 20\,km^2$ (or < 5 locations)
Quantitative analysis of exctinction risk	Population viability analysis (Gärdenfors 2000) or any other form of analysis estimating extinction probability

Note:
(1) Extent of Occurrence (EOO) is the geographical range, defined as the area contained within the shortest continuous imaginary boundary that can be drawn to encompass all the known, inferred or projected sites of present occurrence of a taxon.
(2) Area of Occupancy (AOO) is defined as the area, calculated by summing up all grid squares with the mesh size of $2 \times 2\,km$ that are actually occupied by a taxon, excluding cases of vagrancy.

the levels of threat at the national and regional scale in 16 countries. This figure is mostly focused on Europe due to the limited availability of precise data on species distributions and threat levels on other continents. It nonetheless shows that 2 to 4% of the bryophyte species are already extinct in most countries and that a substantial proportion of the flora is threatened in the short-term. Although data are crucially lacking for most tropical areas, reports of the loss of no less than 12% of the moss species in Puerto Rico in the course of the twentieth century and of 50% of the moss species in Mt Santo Tomas (Philippines) within a 50 year period (Ines-Sastre & Tan 1995), are especially an area of concern.

10.1.3 Implementation of threat levels in legislation

In view of the vulnerability of bryophytes to extinction, a logical step beyond the recognition of highly threatened taxa is the implementation of threat levels into the legislation. Although the degree of legal protection remains very low, bryophyte species have increasingly been included in legal texts regulating the collection of selected species (e.g. the Swiss Nature and Landscape Protection Decree, Appendix 2, www.admin.ch/ch/f/rs/451_1/index.html; the Canadian

Box 10.1
The IUCN categories in brief

In the IUCN classification system, species are ranked into seven categories. Species that belong to either the Extinct, Critically Endangered or Endangered categories, can be termed 'threatened' (Box 10.1 Fig. 1). A brief, simplified description of these categories is provided below (see IUCN 2001, for more details).

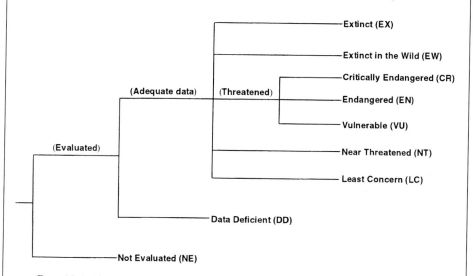

Box 10.1 Fig. 1. Hierarchical diagram of the IUCN threat categories.

Extinct (EX): A taxon is Extinct when there is no reasonable doubt that the last individual has died. For bryophytes, a species is considered EX when it has not been seen for the last 50 years throughout its entire distribution range (EX) or in part of it, in which case the taxon can be included in the Regionally Extinct category (RE). Alternatively, the taxon must be confined to sites, which have been thoroughly surveyed without success in recent years, or where suitable habitat has disappeared.

Critically Endangered (CR) corresponds to a $\geq 50\%$ risk of extinction in 10 years or 3 generations. For bryophytes, the criterion based on the distribution area (AOO, see Table 10.1 for details) is most easy to use. The taxon must be known from a single or several severely fragmented locations* covering altogether $< 10\,km^2$ and experience a continuing decline. Decline in habitat size or quality is most often used as a surrogate for actual population decline in the abscence of data on population trends.

Endangered (EN) corresponds to a risk of extinction $\geq 20\%$ in 20 years or 5 generations. The AOO must be $< 500\,km^2$. In addition, the taxon must be known from ≤ 5 locations* and also have experienced a continuing decline.

Box 10.1 (cont.)

Vulnerable (VU) corresponds to a risk of extinction ≥ 10% in 100 years. The AOO must be < 2000 km². In addition, the taxon must be known from ≤ 10 locations* and have experienced a continuing decline.

Data Deficient (DD) is applied to taxa with insufficient data to categorize them, but which are thought likely to qualify as Extinct, Critically Endangered, Endangered or Vulnerable when they are better known.

Near Threatened (NT): A taxon is NT when it is close to qualifying for VU.

Least Concern (LC) is applied to taxa that do not qualify (and are not close to qualifying) as threatened or near threatened.

Note:

*The term 'location' defines a geographically or ecologically distinct area in which a single threatening event can rapidly affect all individuals of the taxon present (IUCN 2001).

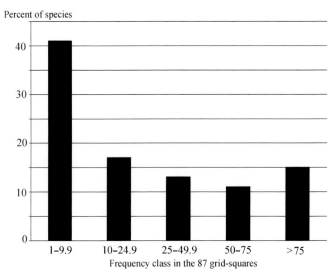

Fig. 10.1. Bryophyte species frequency distribution in 87 4 × 4 km grid-squares in an area of southern Belgium, showing a typical skewed-shape (reproduced from Vanderpoorten & Engels 2003 with permission of Springer).

Species at Risk Act (McIntosh & Miles 2005); the European Annex V of the Habitats Directive, ec.europa.eu/environment/nature/nature_conservation/ eu_nature_legislation/habitats_directive/index_en.htm; and Schedule 8 of the British Wildlife and Countryside Act 1981, www.jncc.gov.uk/). In the European Union for example, the habitats of 32 threatened bryophyte species are protected under Annex II of the Habitats Directive, which has already led

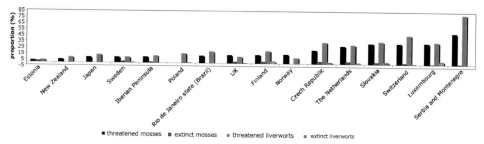

Fig. 10.2. Proportion of extinct and threatened (sensu IUCN) moss and liverwort species in the bryophyte flora of selected countries and areas. Threatened species correspond to the categories 'critically endangered', 'endangered' and 'vulnerable' of IUCN (Box 10.1) (data from Vanderpoorten & Hallingbäck 2009).

to the protection of more than 1000 localities included in the Natura 2000 network (Hallingbäck 2003). Similarly, in the UK, bryophytes may be included within a Site of Special Scientific Interest (SSSI) and guidelines have been produced to aid the selection of SSSI on the basis of the sum of species threat levels, as defined in the British Red Data Book (www.jncc.gov.uk/).

10.2 Why are bryophytes threatened?

10.2.1 Direct threats

The high water-holding capacity of bryophytes, especially those with highly specialized hyalocysts such as *Leucobryum* and *Sphagnum*, makes them a useful potting medium, particularly favoured by orchid growers and for wrapping flowers or fruit tree rootstocks for transportation (Box 2.1). At present, although outdoor *Sphagnum* nurseries are an interesting option for a new type of professional horticulture (Rochefort & Lode 2006), bryophytes are directly harvested from natural populations. Local regulations sometimes exist (for example, the EU Directive 92/43/EEC (Habitats Directive) in Europe, eur-lex.europa.eu/LexUriServ/LexUriServ.do?uri=CELEX:52001 DC0162(02):EN:HTML; and the Flora and Fauna Guarantee in Victoria, Australia), but this activity is not monitored globally and can result in considerable ecological damage and decline in bryophyte diversity. Initially mostly focused on *Sphagnum* (e.g. Whinam & Buxton 1997), commercial moss harvest has been increasing and expanding to other taxa, including epiphytes. Epiphytic moss harvest is most developed in temperate rain forests, such as those found along the Pacific coast of western North America. Commercial harvesting is a sensitive issue in this area, which appears to be one of the most important regions for bryophyte conservation worldwide because 15% of the moss species are endemic (Schofield 1984). Long-term evaluation of cover and species richness based on simulated commercial moss harvest indicates that cover regrowth may require at least 20 years and biomass recovery even longer (Peck 2006). Slow rates of accumulation and the unwanted harvest of non-target species (including red-listed species), provide the incentive to manage and monitor the harvest in order to ensure sustainability and maintain diversity (Muir *et al.* 2006, Peck 2006, Peck & Christy 2006).

10.2.2 *Indirect threats*

While bryophytes may be directly threatened through harvest in specific areas for reasons detailed above, most serious threats originate indirectly from the destruction or degradation of their habitats. The causes are many and operate at both the global and local scales. At the global scale, threats include factors such as global warming, pollution and biological invasions. Invasive plants often constitute a threat to biodiversity and have consequently received much attention recently. Surveys of the vegetation of permanent plots in Dutch sand dunes demonstrated, for example, that *Campylopus introflexus* has continuously increased in abundance since the 1960s, outgrowing the native vegetation and finally forming a monotonous, dense moss carpet with an open canopy of the grass *Corynephorus canescens* (Biermann & Daniels 1997, Ketner-Oostra & Sykora 2004). At the local scale, deforestation, for both agriculture and timber harvest, at rates as high as 2.9 million ha per year in areas such as Brazilian Amazonia, is one of the most serious areas of concern (Laurance 1998). Such habitats, which were once continuous, have become increasingly fragmented. After intensive clearing, the fragments tend to be very small and isolated from each other by crop land, pasture, pavement, or even barren land. This process of fragmentation is one of the main issues in conservation (Heywood & Iriondo 2003).

In fact, it has been widely acknowledged, since McArthur and Wilson's (1967) equilibrium theory of island biogeography, that the number of species reflects a balance between immigration and extinction rates, which are both influenced by two main intrinsic island characteristics. On the one hand, island size impacts on immigration and extinction by influencing the likelihood of diaspore deposition and the availability and duration of suitable habitats. On the other hand, immigration is controlled by species dispersability and thus by the distance separating areas to be colonized and source populations. As a result, species diversity is expected to be correlated with both island size and connectivity with continental source areas. This conceptual model has since been widely applied to predict species richness in both oceanic islands and terrestrial habitats (Whittaker 1998).

Increased extinction risk in isolates results from interactions between local and regional population dynamics. At the local scale, small fragments exhibit proportionally more degraded edge surfaces than large ones. Edges are characterized by high wind turbulence, light levels, rates of evaporation, ambient temperature and low humidity. These factors primarily impact on epiphytes. Epiphytic edge communities can be as rich or even richer than interior, undisturbed communities (Vanderpoorten *et al.* 2004a, Holz & Gradstein 2005), but the composition of their epiphytic assemblages markedly differs. In fact, epiphytes can be classified into 'sun' and 'shade' species. Shade epiphytes are characteristic of the understorey of dense primary forests (see Fig. 10.3). They are less desiccation-tolerant than sun epiphytes (Section 8.1.2) and generalists that developed a series of putative adaptations, such as papillose cell walls, which enhance the capillary absorption and speed up the process of rehydration (Section 4.1). Shade epiphytes are, therefore, highly sensitive to disturbance (Gradstein 1992a, b, Gradstein *et al.* 2001, Acebey *et al.* 2003) and the ratio of shade versus light epiphytes is used as a surrogate of 'naturalness' of forest stands (Box 10.2).

At the regional scale, increased insularity associated with fragmentation potentially reduces migration rates among isolated populations relative

Fig. 10.3. The laurel forests of the Canary Islands, Madeira and the Azores are the relicts of a sub-tropical forest that spread across Europe during the Tertiary Period and was decimated during the glaciations. Those of

Caption for Fig. 10.3. (cont.)
Anaga on Tenerife (a) and the Garajonay National Park on La Gomera (b)
(photo V. Hutsemékers), a UNESCO World Heritage Site, are noteworthy
for their rich bryophyte assemblages (González-Mancebo *et al.* 2004),
including numerous endemics, such as the moss *Echinodium prolixum* (c)
(photo J.-P. Frahm), which forms lush 'curtains' hanging on the branches
of the laurels. See plate section for colour version.

Box 10.2
Epiphytic bryophytes as biodiversity indicators

The parameters defining the conservation value of forest stands are numerous and complex (e.g. tree composition, regeneration mode and structure, amount of dead wood and overmature trees). Biological indicators are therefore frequently used for assessing the conservation value in such complex ecosystems. Epiphyte bryophytes exhibit a series of features that make them ideal bioindicators of disturbance in forests. Because bryophytes are poikilohydric, their physiology is directly controlled by ambient air humidity (Section 8.1). In particular, shade epiphytes, which are characteristic of the understorey of dense primary forests, are considered to be less drought resistant than sun epiphytes and generalists. Shade epiphytes are therefore highly sensitive to deforestation (Gradstein 1992a, b, Gradstein *et al.* 2001, Acebey *et al.* 2003) and more likely threatened (Gradstein *et al.* 2001).

Furthermore, demographic processes are especially crucial in fugitive species such as epiphytes, for which substrate availability lasts for a limited amount of time. Thus, epiphytes with a low dispersal ability are most sensitive to disturbance because they do not have the capacity to respond quickly to a change in their environment (Boudreault *et al.* 2000, Cobb *et al.* 2001, Berg *et al.* 2002). Therefore, large old trees, which are available for a longer time for colonization, tend to harbour more species with low colonization abilities than young ones.

It is precisely this low resilience of shade epiphytes that is exploited for bioindication. For example, Drehwald (2005) presented a biomonitoring system of disturbance in tropical rainforests based on floristic studies, wherein indicators for primary undisturbed forests versus secondary forests and isolated trees are distinguished. Indeed, although primary and secondary forests display similar diversity patterns because harvested areas are rapidly invaded by sun epiphytes (Hyvönen *et al.* 1987), the composition of their epiphytic assemblages markedly differ. One third of primary forest species had not re-established in Costa Rican secondary forests after 40 years of succession, indicating that a long time is needed for the re-establishment of microhabitats and reinvasion of species and communities adapted to differentiated niches (Holz & Gradstein 2005). Bioindicators are therefore selected so that (i) they exhibit a narrow habitat specificity; (ii) they are taxonomically not problematic; (iii) their identification, often to the genus level only, is relatively easy; (iv) they are described in recent floras. The method involves a simple protocol, according to which the occurrence of indicative taxa at < 2 m from 10–20 trunks exhibiting well-developed bryophyte vegetation is recorded. From the list of taxa and their indicator value, an index of 'naturalness' can be derived (Box 10.2 Fig. 1).

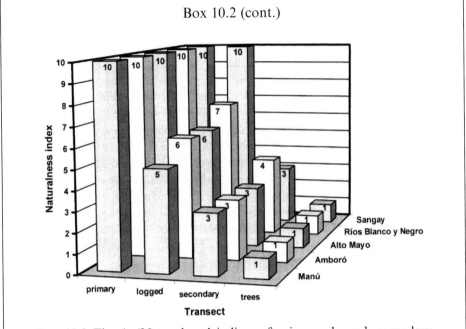

Box 10.2 Fig. 1. 'Naturalness' indices of primary, logged, secondary Neotropical forests and isolated trees derived from their composition in indicator epiphytic bryophytes at five low montane and lowland rainforest areas in South America (reproduced from Drehwald 2005 with permission of the Hattori Botanical Laboratory).

to extinction rates, resulting in a non-equilibrium metapopulation dynamic (Hanski 1999). For example, Zartman and Shaw (2006) measured population size biannually over four years and tallied the number of colonization and extinction events of two epiphyllous liverworts on the leaves of 98 palm trees within an experimental design of lowland Amazonian forest patches of 1, 10 and 100 to 110 000 ha. They measured the rate of colonization as the fraction of initially vacant palm leaves subsequently colonized by either of the two focal epiphyll species within the four years of the experiment. Leaf patches in forest stands of > 100 ha experienced nearly twice the colonization rates observed in small stands (Fig. 10.4), suggesting that the cause of epiphyll species loss in small fragments (< 10 ha) (Fig. 10.5) is reduced colonization.

In the long-term, fragmentation can also have deleterious effects on levels of genetic diversity in impacted populations. Fragmentation is expected to result in a loss of genetic diversity in small, isolated populations by the random process of genetic drift. In a study on the impact of peatland

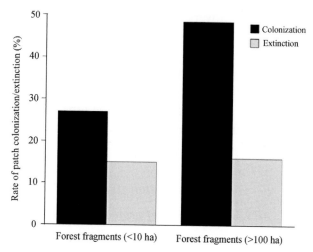

Fig. 10.4. Colonization and extinction rates measured for populations of two epiphyll liverwort species on the leaves of 98 palm trees within an experimental design of lowland Amazonian forest patches of < 10 and > 100 ha. Population size and number of colonization and extinction events were recorded biannually over four years. Rate of colonization is defined as the fraction of initially vacant palm leaves subsequently colonized by either of the two focal epiphyll species within the four years of the experiment (redrawn from Zartman & Shaw 2006).

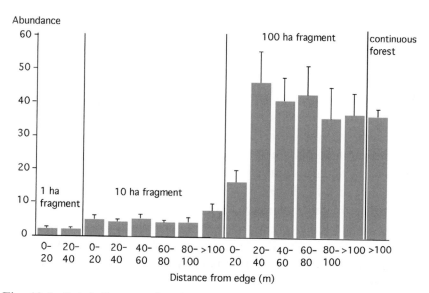

Fig. 10.5. Epiphyll mean abundances ± SD (i.e. number of bushes <2 m having ≥1 epiphyll colonies in 20 × 20 m plots) in 1 to 100 ha forest fragments and continuous forest as a function of proximity to forest border at the Biological Dynamics of Forest Fragments Project in Manaus, Brazil (reproduced from Zartman & Nascimento 2006 with permission of Elsevier).

fragmentation on the genetic diversity and structure of the peat moss *Polytrichum commune*, genetic diversity values from completely cut bogs were indeed found to be lower than those from uncut peatlands (Wilson & Provan 2003). Other studies, however, failed to demonstrate a significant impact of fragmentation on population genetic diversity and structure in *Sphagnum* (Thingsgaard 2001). These contrasting results suggest that the long-term consequences of fragmentation may be differently expressed in different taxa with contrasting life history strategies and dispersal ability.

10.3 Conservation strategies

10.3.1 Selection of protected areas

Biodiversity conservation has traditionally relied on the establishment and maintenance of a network of protected areas. In view of the impossibility of accurately documenting global biodiversity patterns, conservation criteria are mostly based on flagship taxa. For bryophytes in particular, conservation is mainly incidental and occurs through preservation of habitat for other reasons (Cleavitt 2005). How effectively diversity in general, and bryophyte diversity in particular, is currently incorporated in the system, is generally not known (Andelman & Fagan 2000, Pärtel *et al.* 2004, Schulze *et al.* 2004).

The diversity of one group of organisms can, in fact, hardly be predicted by the diversity of another group (Schulze *et al.* 2004, Norden *et al.* 2007) and many groups of organisms actually differ in their conservation demands (Heilman-Clausen *et al.* 2005). In Chile, species richness patterns for vascular and non-vascular plants are inverted across 25 degrees of latitude (Rozzi *et al.* 2008). At a local scale, Pharo *et al.* (2000) found that a set of eastern Australian forest sites that preserved 90% of vascular plant species captured only 65% of bryophyte species and that vascular plant species richness was a poor predictor of bryophyte species richness. Reserves selected for vascular plants may incidentally capture large percentages of bryophytes, but some sites of high bryophyte diversity would not be a conservation priority based on their vascular plant diversity. Many habitats, such as shaded rock outcrops, crags, old quarries, waterfalls and so on, which are not interesting for their vascular flora, often display a very rich bryoflora. Therefore, some habitats should be specifically protected with regard to their bryophyte flora.

10.3.2 Management of protected areas

Once selected, protected areas must be managed to ensure the long-term viability of species or communities that justified the conservation status of these areas. The vegetation dynamics and sometimes rapid species turnover, however, raise such issues as to which state or stage of the vegetation cycle should be preserved (Heywood & Iriondo 2003). Therefore, the contradictory goal of conserving a biota that is dynamic and ever-changing can only be solved when appropriate temporal and spatial scales are set. The approach to protected areas has in fact considerably changed during the last 20 years. The 'fortress' concept, which dominated conservation philosophy in earlier decades, has progressively moved towards a much more interventionist approach involving the acceptance of a broad range of options and techniques (Marrero-Gomez *et al.* 2003).

The most appropriate actions for recovering declining populations can be determined by experiments that test the effects of different management regimes. Except for Scandinavian countries, where cryptogams are often taken into account in management plans for the boreal forests, reserves are almost never managed for bryophytes. Rather, the landscape is mostly managed in favour of a suite of species perceived to be the most sensitive ('focal-species' approach, Lambeck 1997). The fundamental assumption is that if restoration efforts are targeted towards a group of species, the needs of other taxa will also be met. However, some authors have raised concerns about the conceptual, theroetical and practical basis of taxon-based surrogate schemes (Andelman & Fagan 2000, Lindenmayer *et al.* 2000, 2002).

For example, most temperate dry grasslands are a man-made, semi-natural vegetation type of high biological value that has been threatened due to the loss of its agricultural usefulness in the middle of the twentieth century. Many of them were therefore set aside as nature reserves, which have to be actively managed to prevent a natural succession to woodland. The conservation of biological diversity in grasslands requires an integrated approach covering the ecological demands of a multitude of organisms. In practice, however, the emphasis is often placed on the vascular flora. Bryophytes, which include a number of rare species restricted to that habitat (During 1990), are seldom considered in conservation and restoration programmes and the extent to which management practices affect the bryophyte layer are largely unknown.

In calcareous grasslands, cessation of management results in the development of a tall and dense herb and shrub canopy that eventually causes bryophyte diversity to decrease (van Tooren *et al.* 1991). In order to conserve the bryophyte species richness that is specific to the grassland vegetation, a certain amount of disturbance to prevent shrub recolonization is needed. Mowing is often implemented because of its positive effects for orchids. For bryophytes, however, mowing is not an optimal management strategy

(Bergamini *et al.* 2001b). Vanderpoorten *et al.* (2004b) observed that mown plots are characterized by a tight bryophyte carpet mostly composed of the large *Pseudoscleropodium purum*. The latter is a strong competitor, which is likely to outcompete typical bryophyte grassland species. Alternatively, because bryophyte richness is inversely related to graminoid abundance (Yates *et al.* 2000, Klanderud & Totland 2005, Pharo *et al.* 2005, Eskelinen & Oksanen 2006), grazing, which opens the moss and grass layers, is likely to increase species richness, especially that of gap-detecting colonists (van Tooren *et al.* 1990).

10.3.3 *Reintroduction and* ex situ *conservation*

Reintroduction is a particular type of management focused on the conservation of a single target species. For species with a low dispersal range, reintroductions are justified when the widening of the gap between source populations and potential habitats constitutes a major impediment to local and regional demographics. Hence, for dispersal-limited species, conservation efforts to maintain a viable number of populations in the long-term may require artificially introducing diaspores into new suitable localities, or to old localities from which the species has disappeared. Gunnarsson and Söderström (2007) documented a case of reintroduction of *Sphagnum anger-manicum* in Sweden. Two pieces of evidence suggest that the rarity of the species originated from a dispersal limitation rather than a globally compromised habitat quality. First, the species is absent from many suitable habitats. Second, the species never produces sporophytes. To test this hypothesis, Gunnarsson and Söderström (2007) harvested capitula from source populations and transplanted them into suitable, but unoccupied sites. In fact, the technique proved to be useful to successfully create new populations or increase the population size of already existing populations.

In the case of *S. angermanicum*, all Swedish populations appear to be genetically or morphologically quite similar, so that the choice of source populations may be fairly unimportant given the lack of evidence of local adaptation (Gunnarsson *et al.* 2005). Bryophytes in general, however, display similar amounts of intra-specific genetic variation as angiosperms (Hedderson & Nowell 2006), although such variation may often be hidden under a constant morphology (Shaw 2001). In *Sphagnum*, in particular, micro-evolutionary processes leading to local adaptation have been demonstrated among populations from different habitats (Såstad *et al.* 1999). A minimal precaution in any attempt of reintroduction therefore involves a detailed genetic study of the source populations in order to ensure that the appropriate genetic strains are used at the right place.

When populations of the species to be reintroduced do not exist in the wild anymore, reintroductions may involve the use of specimens that have been conserved *ex situ*. *Ex situ* conservation involves the medium- or long-term storage of selected samples of a population's genetic diversity intended for the possible reintroduction of rare and endangered taxa into the wild.

Typical *ex situ* techniques involve the creation of living collections or diapaused material such as seeds, spores or cryopreserved material. Living collections can be grown in greenhouses or axenic cultures (in Petri dishes). Axenic cultures, although artificial, provide a uniform and secure method of maintaining plant tissues without fungal, algal and bacterial contaminants (Duckett *et al.* 2004). Rothero *et al.* (2006) documented a case of reintroduction of the rare spring moss *Bryum schleicheri* var. *latifolium* from axenic cultures. Plants collected from the only UK locality of the species were successfully grown on a phytagel medium. Within 3–4 weeks of initiating cultures from stem fragments, the surface of the medium in 5 cm Petri dishes was completely covered with protonema bearing new shoots (Fig. 10.6a). The cultures were subsequently used for reintroduction. Muslin-encased cultures were pegged down in areas where the vegetation had been cleared. Four out of the 21 introduced colonies survived a full year and developed new shoots (Fig. 10.6b, c). Further investigations are obviously required to determine optimal conditions for transplantation and a monitoring survey remains necessary to assess the long-term success of the reintroduced plants. Nevertheless, this attempt suggests that the reintroduction of bryophytes from cultures might be a promising conservation technique in the near future.

Axenic cultures, however, might not be ideal for long-term conservation purposes. Indeed, the material is continually sub-cultured and is likely to become adapted to growing in culture conditions over time. This is particularly problematic for material retained for conservation purposes where reintroduction is a possible long-term objective. Alternatively, cryopreservation, which consists of the preservation of cells or tissues at $-196°C$ in liquid nitrogen, has been shown to be an effective, convenient and stable long-term storage technique. In bryophytes, the first experiments appear promising, although desiccation-intolerant species did not survive either dehydration or freezing (Burch 2003). A project for the *ex situ* conservation of endangered UK bryophytes was launched in August 2000, at the Royal Botanic Gardens, Kew, with the appointment of a dedicated bryophyte conservation officer.

Fig. 10.6. *In vitro* cultivation and reintroduction into the wild of rare and threatened bryophytes species: the example of the moss *Bryum schleicheri* var. *latifolium* in the UK. (a) Phytagel culture in 5 cm Petri dishes. (b) Colony on a muslin bag six months after planting. (c) A well-grown colony one year after planting (photos G. P. Rothero and J. G. Duckett). See plate section for colour version.

The aim is to provide long-term basal storage of rare bryophyte material for use in future conservation programmes and material of 18 endangered species is currently conserved (see further details at rbg-web2.rbge.org.uk/bbs/Learning/exsitu/exsitu.htm).

10.4 Restoration ecology

When habitats have been destroyed or damaged to such an extent that they cannot recover by themselves, restoration actions are necessary. Ecological restoration is the process of assisting the recovery of an ecosystem that has been damaged, degraded or destroyed. Peatlands are the major bryophyte habitats in need of restoration, since peat extraction commonly involves the drainage of large *Sphagnum*-dominated areas (Box 2.1). During this process, the vegetation layer is removed and the peat is milled before organic matter is harvested with large vacuums or tractors. Drainage and removal of surface vegetation and soil layers alters soil structure and water balance. Increased peat subsidence due to, for example, compression and shrinkage, decreases peat pore size, reducing the water storage capacity (Turetsky & St Louis 2006).

The goal of restoration in peatlands is to re-establish both a plant cover dominated by *Sphagnum* and the functions of the ecosystem, including accumulation of carbon. Restoration practices therefore tend to shorten the successional pathway towards plant communities that are functionally similar, if not structurally similar, to the pre-disturbance state (Campbell *et al.* 2003).

10.4.1 Why are restoration actions necessary?

The process of recolonization is slow. Many typical species, especially *Sphagnum*, are absent from mined sites and are unlikely to recolonize these naturally (Salonen & Setälä 1992, Salonen *et al.* 1992, Whinam *et al.* 2001). Mined peatlands are primarily reinvaded by birch (Lavoie & Rochefort 1996). The establishment of a birch or poplar woodland and the lack of a well-developed *Sphagnum* carpet are not conducive to the rapid recovery of a functioning peat-accumulating system (Rochefort 2000).The flat, extensive mined peat fields thus tend not to regenerate naturally into a functioning peatland, at least not within the first 25 years (Rochefort & Lode 2006). Both dispersal limitations and ecological constraints account for the failure of mined peatlands to regenerate naturally into functional accumulating bogs.

10.4.1.1 Dispersal limitations for recolonization

Because of the absence of diaspore banks (Box 7.3) in peat bogs, recolonization of mined sites must originate from immigrant diaspores. The immigration potential of prospective colonists depends on their residual occurrence in edges, their fecundity and their dispersal ability. In a study on the colonization potential of bryophyte and angiosperm species in mined areas in Quebec, Campbell *et al.* (2003) found that mosses, especially *Polytrichum strictum* and *Pohlia nutans*, were by far the most fecund species (Fig. 10.7). They also displayed a high potential for regeneration from vegetative fragments. Mosses, furthermore, exhibit, despite a low release height, a high wind dispersal ability thanks to their very low wing loadings (ratio of diaspore mass to area) and long fall times as a result of the small size of the spores. Of all the investigated peat bog species, *P. strictum* showed the highest overall immigration potential, resulting from moderate frequency in edges, high

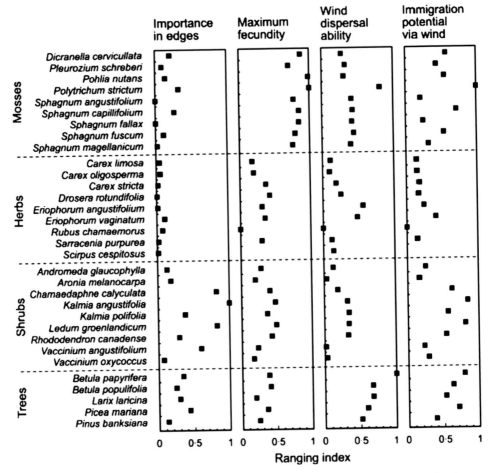

Fig. 10.7. Colonization potential of bryophytes and angiosperms in mined areas in Quebec. The overall immigration potential of a species is a function of (i) its frequency in the edges; (ii) its fecundity, defined as the product of (a) the density of fruiting bodies in quadrats of 50 × 50 cm for angiosperms and 25 × 25 cm for mosses over areas of high concentration in selected peatlands of southeastern Quebec and (b) the average number of seeds/spores in each fruiting body; and (iii) its wind dispersal ability, which results from the release height of the diaspores, fall time and propagule wing loading (mass/area) (reproduced from Campbell *et al.* 2003 with permission of Blackwell).

spore production and high wind dispersal ability (Fig. 10.7). Other mosses exhibited a relatively high immigration potential because of their fecundity and dispersal ability.

Despite their high immigration potential, mosses are substantially under-represented on abandoned milled peatlands compared with angiosperms (Fig. 10.8). The discrepancies between immigration potential and actual

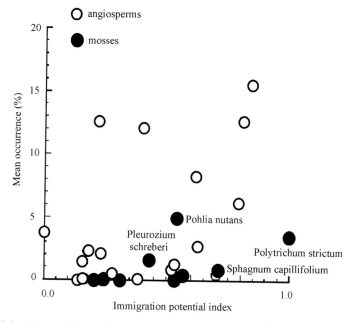

Fig. 10.8. Comparison of the mean occurrence of selected angiosperm and moss species in abandoned milled peatlands in Quebec as a function of immigration potential by wind dispersal (reproduced from Campbell *et al.* 2003 with permission of Blackwell).

colonization suggest that ecological factors control their colonization success after immigration, which we examine in the next section.

10.4.1.2 Unsuitability of ecological conditions in mined peat bogs for recolonization

The bare peat of mined areas represents an exceedingly harsh habitat for both hydrological and climatic reasons. Following mining, the decomposition of the peat accelerates owing to increased mineralization and its water permeability properties are lower. As a consequence, the water table sinks far below the peat surface. The surface layer consequently dries up and forms a crust that is inadequate for diaspore germination. In fact, in the absence of the thick mat of dead sphagna that composes the upper layer of the bog in healthy peatlands, the water pressure strongly increases, making water unavailable to sphagna.

In addition, the extreme microclimate conditions of bare peat severely prevent successful establishment. In the winter, frost heaving may raise a crust of soil by several centimetres in a single night and generate unstable conditions that prevent the successful establishment of peat mosses. In the

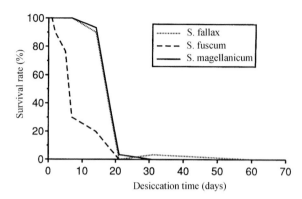

Fig. 10.9. Effect of length of desiccation period on *Sphagnum* survival, expressed as the percentage of 30 stems that showed signs of regeneration after 10 weeks on a water-saturated medium (reproduced from Sagot & Rochefort 1996 with permission of *Cryptogamie, Bryologie*).

summer, recurrent periods of aridity occur and surface temperatures can reach up to 70°C. *Sphagnum* species are especially vulnerable to drought and substrate instability during the establisment phase (Rochefort 2000). They can survive several days without rain, but only if air relative humidity is sufficient. In an experiment on desiccation tolerance, *S. fallax* and *S. magellanicum* exhibited almost 100% survival after 8 days at 20°C and 60% air relative humidity. By contrast, survival was only 50% in more sensitive species such as *S. fuscum*. All species died after 21 days (Fig. 10.9). At 30–40°C, most species could not survive two days of desiccation (Sagot & Rochefort 1996). These results suggest that extreme climate conditions of bare peat severely prevent successful establishment.

10.4.2 Modalities of peatland restoration

Because spontaneous regeneration does not take place as a result of the deep alteration of the habitat in mined peat bogs, restoration actions need to be implemented in order to increase establishment rates by improving habitat conditions. The first years of trials to reintroduce sphagna on bare peat (late 1980s and early 1990s) by simply spreading fragments on bare peat failed to regenerate the moss cover. Successful peatland restoration requires a considerably more complicated process. In fact, four management interventions are strictly required for a successful restoration: field surface preparation; active reintroduction of shoot fragments; application of a protective mulch cover; and rewetting by blocking drainage (Rochefort & Lode 2006).

Fig. 10.10. Successive mechanical steps for successful restoration of milled harvested peatlands. (a) Crust removal. (b) Diaspore collection at a donor site. (c) Shredding and spreading of diaspores onto the area to be restored. (d) Straw mulch application. (e) Ditch blocking (reproduced with permission of the Peatland Ecology Research group). See plate section for colour version.

Surface preparation consists of the breaking up of the crust (Fig. 10.10a). This operation allows better contact between shoot fragments and peat surface as well as greater access to soil moisture. Blocking of the drainage system (Fig. 10.10e) is also an absolute prerequiste, although it often takes place at the end of the whole procedure to facilitate the movement of machinery over the site until that point.

Spreading of shoot fragments is necessary for species with low immigration potential. The introduction of species with good immigration potential may

Conservation biology

Fig. 10.11. Outcome of a peatland restoration programme in eastern Canada. (a) Bare peat prior to restoration in 1999, with 71% of bare peat.

still be required, though, as a result of the large size of milled peatlands. Plants are collected from nearby bogs. The operation consists in picking up the first 6–10 cm of the vegetation (Fig. 10.10b) in a quantity that is proportional to the area of the site to be restored, usually in a proportion of about 1:10–1:15. Subsequent shredding to a size of 1–3 cm allows better spreading and increases the regeneration rate as small fragments make better contact with the substrate (Fig. 10.10c).

Finally, shoot fragments need to be protected against erosion and desiccation. This can be accomplished by the complementary operations of mulching and the use of companion species. The use of natural or artificial mulch decreases thermal variation of soil and reduces evaporation (Fig. 10.10d) (Price *et al.* 2003). Straw mulch, which is most commonly used, is, however, progressively blown away. To complement mulching, the use of pioneer, companion species such as *Polytrichum* can improve the microclimate over a cutover peat surface, reduce the negative effects of frost heaving, and hence substantially contribute to the success of the restoration (Groeneveld & Rochefort 2005). The extent to which such 'nurse plants' enhance establishment of *Sphagnum* or eventually negate the facilitation through competition with the *Sphagnum* remains to be tested.

This suite of restoration actions is likely to revegetate a cutover peatland with peatland plants and stabilize the peat within 3 to 5 years (Rochefort & Lode 2006). In a restoration experiment of a mined peat bog abandoned 20 years ago with an initial proportion of 71% of bare peat (Fig. 10.11a), moss carpets covered 62% of the surface after 2 years, of which 22% was *Sphagnum* (Fig. 10.11b). Four years after restoration began, sphagna covered 35% of the surface (Fig. 10.11c). One year later, *Polytrichum* and *Sphagnum* carpets covered 70% of the ground. By contrast, *Sphagnum* still covered <0.2% in the control site (Rochefort & Lode 2006).

Caption for Fig. 10.11. (cont.)
(b) In 2001, two years after restoration following the procedure illustrated in Fig. 10.10, the same area with 62% moss carpet cover, of which 22% was *Sphagnum*. (c) In 2004, with 80% of total moss cover (reproduced with permission of the Peatland Ecology Research group). See plate section for colour version.

Glossary

ABA
abbreviation for abscisic acid, a phytohormone that acts as a signalling molecule helping to control overall growth and physiology.

acrocarpy
development of a perichaetium at the apex of the primary module (i.e. stem); see also pleurocarpy and cladocarpy. In acrocarpous mosses, vegetative growth of the stem is terminated by the development of apical sex organs. Although 'carpy' refers to the female sex organs, the terminal position may be occupied by antheridia.

acrogyny
initiation, in most leafy liverworts (Jungermanniidae), of the development of an archegonium by the apical cell (compare to anacrogyny). In an acrogynous taxon, growth of the stem is thus terminated by the development of female sex organs and vegetative growth is resumed via lateral branches.

acropetal
refers to a gradient pointing to the apex (an acropetal branching system is a system wherein the youngest branches are developed towards the apex).

allopolyploid
refers to organisms which contain complete chromosome sets from two distinct species after hybridization.

amphigastrium
(pl. amphigastria)
leaves of the lower surface of the stem of leafy liverworts. Amphigastria (also called underleaves) are typically differentiated in size and shape from lateral leaves.

amphithecium
outer tissue of the sporophyte (surrounding the endothecium), which, in the sporangial region, gives rise to the capsule wall of bryophytes and, in *Sphagnum*, *Andreaea* and hornworts, also to the sporogenous tissue.

anacrogyny | development of an archegonium from a lateral rather than apical cell, typical of most simple thalloid liverworts (Metzgeriidae); compare to acrogyny.

analogy (adj. analogous) | similarity of characters due to convergence rather than inheritence and thus lacking a common origin.

anisospory | intrasporangial variation in spore size in some dioicous mosses, with each spore tetrad comprising two large female and two small male spores.

annulus (pl. annuli) | ring of amphithecial cells that are involved in the dehiscence of the capsule and loss of the operculum in mosses. The cells can be highly differentiated (e.g. *Funaria*) or not (e.g. Orthotrichaceae) and may occur in one or more rows and even in more than one layer.

anticlinal | at a right angle to the surface.

antheridium (pl. antheridia) | male gametangium, composed at maturity of a single-layered jacket on a short stalk and including the sperm cells.

apophysis | modified sterile base or neck of the capsule of mosses that is typically much wider (e.g. *Splachnum*), longer (e.g. *Trematodon*) or larger (e.g. *Tayloria*) than the sporogenous urn (also called hypophysis).

archegoniophore | portion of the thallus of complex thalloids derived through the condensation and constriction of lobes bearing archegonia. The result is an umbrella-shaped structure bearing archegonia. Following sexual reproduction, this structure will carry the sporophyte, and is then called a carpocephalum.

archegonium (pl. archegonia) | Female sex organ composed of a stalk, a venter holding a single egg and a neck that is hollow and open at maturity, and provides the gateway for the sperm cell to reach the egg.

arthrodontous | refers to the peristome of the Bryopsida, where the teeth are built from cell wall remnants (i.e. cell-plates) rather than whole cells as in the nematodontous peristome.

autoicy (adj. autoicous) | distribution of both sex organs on a single plant, but on distinct modules.

axenic | pure, completely germ-free.

basipetal | refers to a gradient pointing away from the apex (a basipetal branching system is a system wherein the branches are initiated towards the base of the stem).

biota | the animal and plant life of a given region.

buffering capacity | ability of a water body to resist large changes in pH when an acidic or alkaline solution is added to it.

calyptra (pl. calyptrae)	membranous structure derived at least in part from the modification of the archegonium following fertilization. The calyptra protects the developing embryo.
capsule	spherical or elongate structure enclosing the spore-producing tissue.
cilia	narrow endostomial appendages developed in-between the broader segments in most mosses with a diplolepideous alternate peristome.
cladocarpy	development of perichaetia, clusters of female gametangia, at the apex of vegetative branches.
cleistocarpy (adj. cleistocarpous)	indehiscence of the sporangium; the capsule lacks an operculum or valves. Spore dispersal follows decay or breakdown of the sporangial wall.
columella	axial sterile tissue in mosses and hornworts surrounded by sporogenous tissue. In some mosses (i.e. *Sphagnum* and *Andreaea*), the columella is short and overarched by the spore sac.
compensation point	point reached when the rate of photosynthesis exactly matches the rate of respiration, i.e. when net photosynthesis is zero.
cucullate	in the form of a hood; used to describe calyptrae that have a longitudinal slit and hence resemble a hood rather than a hat.
cuticle	thin, water-repellent waxy layer on the outer wall of epidermal cells consisting primarily of wax and cutin.
dehiscence	opening of the sporangium or sporophyte to expose the spores that can than be dispersed. In hornworts and liverworts, the dehiscence of the sporangium typically follows longitudinal lines, whereas in most mosses (except *Takakia*, *Andreaea* and *Andreaeobryum*), dehiscence is transversal, leading to the shedding of a lid (operculum).
diaspore	any dispersal device, including spores, gemmae, parts of plants, etc.
dioicy (adj. dioicous)	condition where the antheridia and the archegonia are produced by distinct individuals (i.e. each individual is unisexual).
diplolepideous	refers to a peristome typically composed of both external (exostome) and internal (endostome) layers. The name refers to the architecture of each exostome tooth, built from remnants of two columns of outer peristomial layer cells.
ecological niche	ensemble of environmental conditions that allows a species to survive and reproduce.

ectohydric	with water movement taking place externally.
elaiosome	fleshy outgrowth on a seed with oily substances attractive to ants.
elater	single elongated cell in the sporangium of liverworts, and hence of endothecial origin, with a wall that is reinforced by a spiral thickening, which accounts for the hygroscopic movement of the cell upon dehydration (see also pseudo-elaters).
embryo	first multicellular stage in the life cycle of land plants. It is formed via mitotic divisions from the zygote and develops into the sporophyte.
embryophytes	plants bearing embryos.
endohydry	refers to the ability to move water (-hydry) internally (endo-) within the plant body.
endospory	premature mitotic division of the spore prior to dispersal, resulting in a multicellular sporeling enclosed in the spore wall.
endostome	inner peristome in diplolepidous or sole ring in haplolepidous mosses. The dorsal surface of the endostome tooth is built from the cell plates derived from the primary peristomial layer, whereas the ventral (inner) surface is built from cell plates from the inner peristomial layer. The endostome of diplolepidous mosses may be composed of segments, cilia and a basal membrane. See also exostome.
endothecium	inner tissue of the sporangium (internal to the amphithecium) that gives rise to the columella (when present) and the sporogenous tissue in all bryophytes except *Sphagnum*, *Andreaea* and hornworts.
epigonium	enlarged archegonial tissue enclosing the fertilized egg.
epilithic	growing on surface of rocks.
epiphragm	thin membrane spanning the mouth of the capsule in some nematodontous mosses (i.e. Polytrichaceae) and bearing small perforations along its circumference.
epiphyll	a specialized epiphyte growing on leaves.
eutrophication	process whereby the environment becomes rich in nutrients through natural (mineralization of organic matter) or man-induced (e.g. inputs of fertilizers) processes.
exospory	germination of the spore resulting in a protonemal phase emerging from the spore.
exostome	outer peristome in diplolepidous mosses. The dorsal surface of the exostome tooth is built from the cell

plates derived from the outer peristomial layer, whereas the ventral (inner) surface is built from cell plates from the primary peristomial layer. See also endostome.

exothecial cells   outer cells of the capsule wall.

feather mosses   a phylogenetically unrelated group of mosses characterized by forming relatively robust, intertwined branching systems and occurring on the forest floor in large abundance. Common in the boreal region are *Hylocomium splendens* and *Pleurozium schreberi*.

Fv/Fm ratio   measure of chlorophyll fluorescence. Part of the light energy accumulated by chlorophyll is restituted as fluorescence (Fig. 8.11). This corresponds to the 'background' fluorescence level F_0 in plants under natural conditions. In plants exposed to stress, chlorophyll tends to degrade and, hence, become less fluorescent. Chlorophyll fluorescence thus provides a sensitive indicator of stress conditions in plants. F_m, termed maximum fluorescence, is a measure of fluorescence intensity when chlorophyll is artificially prevented from restituting light energy by emitting an electron. All the accumulated energy is therefore restituted as fluorescence. F_m is thus a measure of maximum 'excitability' of the chlorophyll molecule. Variable fluorescence, F_v, is equal to F_m less F_0.

gametangium   sex organ borne on the haploid gametophyte
(pl. gametangia)   (see antheridium and archegonium).

gamete   haploid reproductive cell (see antherozoid and egg).

gametophyte   multicellular haploid stage developing the sex organs. In bryophytes, the gametophyte forms the dominant vegetative phase of the life cycle.

gemma (pl. gemmae)   specialized asexual reproductive propagule.

genetic drift   random fluctuation of allelic frequencies resulting, in small populations, in the fixation of one of the allelic forms.

guide cells   large cells filled with a vacuole found in the costa of some mosses.

gymnostomy   lack of peristome teeth around the mouth of the
(adj. gymnostomous)   capsule.

haplolepidous   refers to a peristome composed of an endostome only. The outer side of each tooth is built from a single column of primary peristomial layer cells.

hardening | increased resistance to an environmental stress resulting from a previous exposure to the same or a related stress.

haustorium (pl. haustoria) | structure at the base of the sporophyte that penetrates the gametophyte and absorbs water, nutrients and photosynthates. The haustorium in liverworts is initially a single elongating cell, which undergoes limited or no further division. Its function is confined to the early stages of sporophyte development.

heterocyst | specialized nitrogen-fixing cells formed by some filamentous cyanobacteria.

homologous | refers to features that are inherited by the descendants from a common ancestor. Over long evolutionary periods, two homologous characters may have diverged significantly and perform distinct functions. See also analogous.

homoplasy | a character-state shared by two or several organisms due to convergent evolution and not shared ancestry.

hyalocyst | hyaline cell typically serving to store water.

hydroid | differentiated, highly elongated cells with thin, non-perforated walls in the stem and also sometimes in the leaf costa of most peristomate mosses.

hypophysis | see apophysis.

incubous | leaf arrangement where the forward margin of one leaf overlies the rear margin of the leaf in front (see also succubous).

inner peristomial layer (IPL) | innermost layer of the amphithecium in peristomate mosses, whose outer periclinal walls compose the inner surface of the endostome in arthrodontous mosses (except in *Mittenia*).

involucre | tube of thallus tissue that protects the developing archegonia of some hornworts and thalloid liverworts.

leptoid | specialized cell in the central strand of the stem, the costa and the seta of some mosses involved in the conduction of photosynthate and thus analogous to the sieve cells of angiosperms.

lignin | polymer impregnating the cell walls of water conducting cells in vascular plants.

matrotrophy | physiological maternal care for the offspring through a physical connection called the placenta.

meiosis | two successive nuclear divisions leading to the reduction of the number of chromosomes in

	the cell, from the diploid (2n) mother-cell to the haploid (n) daughter cells.
metamer	set of cells derived from a single derivative cell produced by the apical cell. All metamers derived from one apical cell compose a module.
microtubule	proteinaceous threads of a diameter of ~24 nm and length varying from several micrometres to millimetres that serve as structural components within cells.
mitosis	process during which the duplicate chromosomes divide and separate to form two genetically identical daughter nuclei. This is followed by the physical division of the cell (i.e. cytokinesis).
mitrate	in the shape of a miter; used to describe calyptrae that cover the capsule like a hat.
module	basic architectural unit of the bryophyte plant body derived from a single apical cell. Each module is composed of metamers.
monoicy (adj. monoicous)	condition where both sexes are produced by the same individual, and hence these individuals are hermaphroditic.
monophyletic	refers to a group defined by a single common ancestor and all of its descendants.
monopodial	a branch system that consists of a primary module with lateral subsidiary secondary modules, resulting in a growth form with only one main stem that continues growing every season. See also sympodial.
nematodontous	refers to the peristome of the Polytrichopsida and Tetraphidopsida, where the teeth are built from whole cells compared to only cell plates as in the arthrodontous peristome.
oligotrophic	nutrient-poor.
ombrotrophic	receiving water only from precipitation.
ontogeny	course of growth and development of an individual to maturity.
operculum (pl. operculi)	flat to conical lid of the capsule of mosses and some liverworts that is shed upon dehiscence.
outer peristomial layer (OPL)	third outermost layer of the amphithecium in peristomate mosses, whose inner periclinal walls compose the outer surface of the exostome teeth in diplolepideous mosses.
outgroup	in a cladistic study, a group that is likely to share ancestral traits with the ingroup (or group of taxa under study) because it shares a close common ancestor with the ingroup.

paraphyllium (pl. paraphyllia)	photosynthetic filamentous or foliose cauline appendages in mosses, likely involved in external water conduction.
paraphysis (pl. paraphyses)	sterile, multicellular and usually uniseriate hairs intermixed with the sex organs in most mosses and some liverworts.
parenchyma (adj. parenchymatous)	a tissue made up of relatively unspecialized, isodiametric thin-walled cells.
perianth	unistratose structure surrounding the developing embryo in liverworts. It is typically developed from the fusion of three leaves.
perichaetium (pl. perichaetia)	cluster of archegonia, often with paraphyses, protected by typically modified leaves (i.e. perichaetial leaves).
periclinal	parallel to the surface.
perigonium (pl. perigonia)	cluster of antheridia, often with paraphyses, protected by typically modified leaves (i.e. perigonial leaves).
photosynthetic photon flux density (PPFD)	incident radiation expressed in moles of photosynthetically active ($l = 400$–700 nm) light quanta per unit area and time.
phylogeny	the evolutionary history of a group of organisms, which is represented by a branching diagram showing the ancestral relations among taxa.
placenta	tissue at the interface of the gametophyte and sporophyte, which is characterized by transfer cells on one or both sides that move water and nutrients along their vast surface area resulting from numerous wall ingrowths.
plasmodesmata	very fine thread of cytoplasm that passes through openings in the walls of adjacent cells and forms a living bridge between them.
plesiomorphy	ancestral state of a trait.
pleurocarpy	refers to the development of female sex organs at the apex of a highly reduced lateral branch typically lacking mature vegetative leaves rather than on well-developed branches (cladocarpy) or at the apex of the stem (acrocarpy).
poikilohydry	ability to freely exchange water with the environment and thus to have an intracellular water content that varies passively with the outside water supply and ambient humidity.
polyphyletic	refers to a group that is not defined by a single common and exclusive ancestor (see also monophyletic).
polyploid	an organism with more than two sets of chromosomes in the nucleus of each cell.

polysporangiophyte   lineage of land plants characterized by the occurrence of several spore-producing organs (i.e. sporangia). The multiplication of the number of sporangia results from the ability of the sporophyte to branch.

primary peristomial layer (PPL)   second innermost layer of the amphithecium in peristomate mosses, whose outer periclinal walls compose the inner surface of the exostome teeth in diplolepideous mosses and whose inner periclinal walls contribute to the outer endostomial surface in all mosses (except *Mittenia*).

protonema (pl. protonemata)   first multicellular stage following the germination of a spore or an asexual propagule. The protonema is thus the equivalent, in mosses, to the sporeling of liverworts and hornworts. However, the protonema lasts longer and develops into a thallus in *Andreaea* and *Sphagum* or a filamentous network in other mosses and is able to produce several gametophytes.

protoplasm   fluid living content of a cell, including the cytoplasm and nucleoplasm.

pseudo-elater   uni- or multicellular filamentous structure produced within the sporangium of hornworts. 'Pseudo-' refers to the distinctiveness of these structures from those developed in liverworts (see elaters). Pseudo-elaters are diploid cells derived from the amphithecium.

pseudoparaphyllium (pl. pseudoparaphyllia)   small appendages on the stem of mosses, around the dormant branch primordium.

pseudopodium   gametophytic axis subtending the sporophyte in *Andreaea* and *Sphagnum*. The pseudopodium develops following sexual reproduction and is thus analogous to the seta, which is lacking in these taxa.

pyrenoid   differentiated regions of the chloroplast where starch is synthesized.

radiation   rapid diversification from a single common ancestor.

rhizoid   filamentous anchoring organs of bryophytes. Rhizoids are unicellular in most liverworts and hornworts, and multicellular in mosses.

saturation point   the irradiance at which the maximum rate of photosynthesis is first attained. Curves fitted to light-response data are generally asymptotic, with a maximum at infinite irradiance, so in practice it is necessary to use the irradiance at which some arbitrary fraction (e.g. 95%) of the theoretical maximum photosynthesis is reached.

schizogeny (adj. schizogenous)	development through the separation of adjacent cells along their middle lamellae. The schizogenous cavities in the thallus of hornworts arise through the spacing of cells.
seta (pl. setae)	stalk of the sporophyte terminated by the sporangium in liverworts and mosses.
shoot calyptra (pl. shoot calyptrae)	in leafy liverworts, tissue derived from both the stem and archegonium, protecting the developing embryo.
sporangium (pl. sporangia)	specialized region of the sporophyte of land plants within which spore mother-cells are formed and undergo meiosis to yield haploid spores.
spore	haploid cell produced by meiotic division of sporangial cells.
sporeling	first multicellular stage following the germination of the spore.
sporocyte	spore mother-cell, i.e. a diploid cell in an embryonic sporangium that undergoes meiosis and produces haploid spores.
sporophyte	multicellular stage of the life cycle of plants characterized by a diploid genome (in comparison to haploid genome of the gametophyte) and producing spores in a specialized tissue, the sporangium.
stereid	thick-walled cell supporting the costa or stem in some mosses.
stoma (pl. stomata)	opening formed in between two differentiated epidermal cells (the guard cells).
succubous	leaf arrangement where the rear margin of one leaf overlies the forward margin of the leaf in front (see also incubous).
symbiosis	a relationship between two species that benefits both partners.
sympodial	a branch system that consists of a chain of connected modules of the same hierarchical level. It is characteristic of acrocarpous taxa with stems of determined growth producing subapical innovations that function as new stem. See also monopodial.
synapomorphy	derived character state shared by some, not necessarily all, descendants from a common ancestor.
tomentum (adj. tomentose)	felted woolly covering of rhizoids on the stem of mosses.
tracheophyte	lineage of embryophytes diagnosed by the presence of specialized water conducting cells that are elongated, have tapered ends, are dead at maturity

	and have walls impregnated with lignin. The common name for tracheophytes is vascular plants.
trigones	heavy thickening of the cell wall in the corners of the laminal cells in liverworts.
tuber	underground vegetative propagule.
vitta	longitudinal band of elongated, often thick-walled cells in a leaf of liverworts, resembling a costa but only one cell layer thick.
zygote	first stage of the diploid phase of the life cycle, resulting from the fusion of a sperm cell with an egg.

References

Acebey, A., Gradstein, S. R. & Kromer, T. (2003) Species richness and habitat diversification of bryophytes in submontane rain forest and fallows of Bolivia. *Journal of Tropical Ecology*, **19**, 9–18.

Adams, D. G. & Duggan, P. S. (2008) Cyanobacteria–bryophyte symbioses. *Journal of Experimental Botany*, **59**, 1047–1058.

Ah-Peng, C. & Rausch De Traubenberg, C. R. (2004) Aquatic bryophytes as pollutant accumulators and ecophysiological bioindicators of stress: bibliographic synthesis. *Cryptogamie Bryologie*, **25**, 205–248.

Alpert, P. (2000) The discovery, scope, and puzzle of desiccation tolerance in plants. *Plant Ecology*, **151**, 5–17.

Alpert, P. (2005) The limits and frontiers of desiccation-tolerant life. *Integrative and Comparative Biology*, **45**, 685–695.

Alpert, P. (2006) Constraints of tolerance: why are desiccation-tolerant organisms so small or rare? *Journal of Experimental Biology*, **209**, 1575–1584.

Alpert, P. & Oechel, W. C. (1985) Carbon balance limits microdistribution of *Grimmia laevigata*, a desiccation-tolerant plant. *Ecology*, **66**, 660–669.

Alpert, P. & Oliver, M. J. (2002) Drying without dying. In *Dessication and Survival in Plants: Drying Without Dying*, eds. M. Black & H. W. Pritchard. Wallingford: CAB International, pp. 3–43.

Andelman, S. J. & Fagan, W. F. (2000) Umbrellas and flagships: efficient conservation surrogates or expensive mistakes? *Proceedings of the National Academy of Sciences of the USA*, **97**, 5954–5959.

Ayres, E., van der Wal, R., Sommerkorn, M. & Bardgett, R. (2006) Direct uptake of soil nitrogen by mosses. *Biology Letters*, **2**, 286–288.

Bakken, S. (1993) Effects of simulated acid rain on the morphology, growth and chlorophyll content of *Hylocomium splendens*. *Lindbergia*, **18**, 104–110.

Barkman, J. J. (1958) *Phytosociology and Ecology of Cryptogamic Epiphytes*. Assen: van Gorcum.

Barthlott, W., Fischer, E., Frahm, J. P. & Seine, R. (2000) First experimental evidence for zoophagy in the hepatic *Colura*. *Plant Biology*, **2**, 93–97.

Bates, J. W. (1982) The role of exchangeable calcium in saxicolous calcicole and calcifuge mosses. *New Phytologist*, **90**, 239–252.

Bates, J. W. (1988) The effect of shoot spacing on the growth and branch development of the moss *Rhytidiadelphus triquetrus*. *New Phytologist*, **109**, 499–504.

Bates, J. W. (1989) Growth of *Leucobryum glaucum* cushions in a Berkshire oakwood. *Journal of Bryology*, **15**, 785–791.

Bates, J. W. (1992) Influence of chemical and physical factors on *Quercus* and *Fraxinus* epiphytes at Loch Sunart, western Scotland: a multivariate analysis. *Journal of Ecology*, **80**, 163–179.

Bates, J. W. (1995) A bryophyte flora of Berkshire. *Journal of Bryology*, **18**, 503–620.

Bates, J. W. (1997) Effects of intermittent desiccation on nutrient economy and growth of two ecologically contrasted mosses. *Annals of Botany*, **79**, 299–309.

Bates, J. W. (1998) Is 'life-form' a useful concept in bryophyte ecology? *Oikos*, **82**, 223–237.

Bates, J. W. (2000) Mineral nutrition, substratum ecology, and pollution. In *Bryophyte Biology*, 1st edn, eds. A. J. Shaw & B. Goffinet. Cambridge: Cambridge University Press, pp. 248–311.

Bates, J. W. (2009) Mineral nutrition and substratum ecology. In *Bryophyte Biology*, 2nd edn, eds. B. Goffinet & A. J. Shaw. Cambridge: Cambridge University Press, pp. 299–356.

Bates, J. W. & Duckett, J. G. (2000) On the occurrence of rhizoids in *Scleropodium purum*. *Journal of Bryology*, **22**, 300–302.

Bates, J. W. & Farmer, A. M. (1990) An experimental study of calcium acquisition and its effects on the calcifuge moss *Pleurozium schreberi*. *Annals of Botany*, **65**, 87–96.

Bates, J. W., Proctor, M. C. F., Preston, C. D., Hodgetts, N. G. & Perry, A. R. (1997) Occurrence of epiphytic bryophytes in a 'tetrad' transect across southern Britain. 1. Geographical trends in abundance and evidence of recent change. *Journal of Bryology*, **19**, 685–714.

Bates, J. W., Thompson, K. & Grime, J. P. (2005) Effects of simulated long-term climatic change on the bryophytes of a limestone grassland community. *Global Change Biology*, **11**, 757–769.

Beaugelin-Seiller, K., Baudin, J. P. & Brotter, D. (1994) Use of aquatic mosses for monitoring artificial radionuclides downstream of the nuclear power plant of Bugey (river Rhone, France). *Journal of Environmental Radioactivity*, **24**, 217–233.

Beckett, R. P., Marschall, M. & Laufer, Z. (2005) Hardening enhances photoprotection in the moss *Atrichum androgynum* during rehydration by increasing fast- rather than slow-relaxing quenching. *Journal of Bryology*, **27**, 7–12.

Benscoter, B. W. & Vitt, D. H. (2007) Evaluating feathermoss growth: a challenge to traditional methods and implications for the boreal carbon budget. *Journal of Ecology*, **95**, 151–158.

Berbee, M. L. & Taylor, J. W. (2007) Rhynie chert: a window into a lost world of complex plant-fungus interactions. *New Phytologist*, **174**, 475–479.

Berendse, F. (1999) Implications of increased litter production for plant biodiversity. *Trends in Ecology and Evolution*, **14**, 4–5.

Berendse, F., van Breemen, N., Rydin, H., *et al.* (2001) Raised atmospheric CO_2 levels and increased N deposition cause shifts in plant species composition and production in *Sphagnum* bogs. *Global Change Biology*, **7**, 591–598.

Berg, A., Gärdenfors, U., Hallingbäck, T. & Noren, M. (2002) Habitat preferences of red-listed fungi and bryophytes in woodland key habitats in southern Sweden: analyses of data from a national survey. *Biodiversity and Conservation*, **11**, 1479–1503.

Bergamini, A. & Pauli, D. (2001) Effects of increased nutrient supply on bryophytes in montane calcareous fens. *Journal of Bryology*, **23**, 331–339.

Bergamini, A., Pauli, D., Peintinger, M. & Schmid, B. (2001a) Relationships between productivity, number of shoots and number of species in bryophytes and vascular plants. *Journal of Ecology*, **89**, 920–929.

Bergamini, A., Peintiger, M., Schmid, B. & Urmi, E. (2001b) Effects of management and altitude on bryophyte species diversity and composition in montane calcareous fens. *Flora*, **196**, 180–193.

Berglund, H. & Jonsson, B. G. (2001) Predictability of plant and fungal species richness of old-growth boreal forest islands. *Journal of Vegetation Science*, **12**, 857–866.

Bidartondo, M. I., Bruns, T. D., Weiß, M., Sérgio, C. & Read, D. J. (2003) Specialized cheating of the ectomycorrhizal symbiosis by an epiparasitic liverwort. *Proceedings of the Royal Society of London* B, **270**, 835–842.

Biermann, R. & Daniels, F. J. A. (1997) Changes in a lichen-rich dry sand grassland vegetation with special reference to lichen synusiae and *Campylopus introflexus*. *Phytocoenologia*, **27**, 257–273.

Billings, W. D. & Drew, W. B. (1938) Bark factors affecting the distribution of corticolous bryophytic communities. *American Midland Naturalist*, **20**, 302–330.

Bisang, I. (1992) Hornworts in Switzerland: endangered? *Biological Conservation*, **59**, 145–149.

Bisang, I. (1996) Quantitative analysis of the diaspore bank of bryophytes and ferns in cultivated fields in Switzerland. *Lindbergia*, **21**, 9–20.

Bisang, I. (1998) The occurrence of hornwort populations (Anthocerotales, Anthoceropisa) in the Swiss Plateau: the role of management, weather conditions and soil characteristics. *Lindbergia*, **23**, 94–104.

Bisang, I. & Ehrlen, J. (2002) Reproductive effort and cost of reproduction in female *Dicranum polysetum*. *Bryologist*, **105**, 384–397.

Bobbink, R., Hornung, M. & Roelofs, J. G. M. (1998) The effects of air-borne nitrogen pollutants on species diversity in natural and semi-natural European vegetation. *Journal of Ecology*, **86**, 717–738.

Boisselier-Dubayle, M.-C., Lambourdière, J. & Bischler, H. (2002) Molecular phylogenies support multiple morphological reductions in the liverwort subclass Marchantiidae (Bryophyta). *Molecular Phylogenetics and Evolution*, **24**, 66–77.

Boudier, P. (1988) Différenciation structurale de l'épiderme du sporogone chez *Sphagnum fimbriatum* Wilson. *Annales des Sciences Naturelles, Botanique*, **8**, 143–156.

Boudreault, C., Gauthier, S. & Bergeron, Y. (2000) Epiphytic lichens and bryophytes on *Populus tremuloides* along a chronosequence in the southwestern boreal forest of Quebec, Canada. *Bryologist*, **103**, 725–738.

Bragazza, L., Siffi, C., Iacumin, P. & Gerdol, R. (2007) Mass loss and nutrient release during litter decay in peatland: the role of microbial adaptability to litter chemistry. *Soil Biology & Biochemistry*, **39**, 257–267.

Brown, D. H. & Wells, J. M. (1990) Physiological effects of heavy metals on the moss *Rhytidiadelphus squarrosus*. *Annals of Botany*, **66**, 641–647.

Brown, D. H. & Whitehead, A. (1986) The effect of mercury on the physiology of *Rhytidiadelphus squarrosus* (Hedw.) Warnst. *Journal of Bryology*, **14**, 367–374.

Brundrett, M. C. (2002) Coevolution of roots and mycorrhizas of land plants. *New Phytologist*, **154**, 275–304.

Budke, J. M., Jones, C. S. & Goffinet, B. (2007) Development of the enigmatic peristome of *Timmia megapolitana* (Timmiaceae; Bryophyta). *American Journal of Botany*, **94**, 460–467.

Buitink, J. & Leprince, O. (2004) Glass formation in plant anhydrobiotes: survival in the dry state. *Cryobiology*, **48**, 215–228.

Burch, J. (2003) Some mosses survive cryopreservation without prior pretreatment. *Bryologist*, **106**, 270–277.

Büscher, P., Koedam, N. & Van Speybroeck, D. (1990) Cation-exchange properties and adaptation to soil acidity in bryophytes. *New Phytologist*, **115**, 177–186.

Cairney, J. W. G. (2000) Evolution of mycorrhiza systems. *Naturwissenschaften*, **87**, 467–475.

Campbell, D. R., Rochefort, L. & Lavoie, C. (2003) Determining the immigration potential of plants colonizing disturbed environments: the case of milled peatlands in Quebec. *Journal of Applied Ecology*, **40**, 78–91.

Carafa, A., Duckett, J. G. & Ligrone, R. (2003) Subterranean gametophytic axes in the primitive liverwort *Haplomitrium* harbour a unique type of endophytic association with aseptate fungi. *New Phytologist*, **160**, 185–197.

Carafa, A., Duckett, J. G., Know, J. P. & Ligrone, R. (2005) Distribution of cell wall xylans in bryophytes and tracheophytes: new insights into the basal interrelationships of land plants. *New Phytologist*, **168**, 231–240.

Carballeira, A., Diaz, S., Vazquez, M. D. & Lopez, J. (1998) Inertia and resilience in the responses of the aquatic bryophyte *Fontinalis antipyretica* Hedw. to thermal stress. *Archives of Environmental Contamination and Toxicology*, **34**, 343–349.

Carballeira, A., Fernandez, J. A., Aboal, J. R., Real, C. & Couto, J. A. (2006) Moss: a powerful tool for dioxin monitoring. *Atmospheric Environment*, **40**, 5776–5786.

Caron, J. (2001) La tourbe et les milieux artificiels. In *Ecologie des Tourbières du Québec-Labrador*, eds. S. Payette & L. Rochefort. Québec: Presses de l'Université Laval, pp. 399–410.

Chapman, S., Buttler, A., Francez, A. J., *et al.* (2003) Exploitation of northern peatlands and biodiversity maintenance: a conflict between economy and ecology. *Frontiers in Ecology and Environment*, **1**, 525–532.

Churchill, S. P. (1998) Catalog of Amazonian mosses. *Journal of the Hattori Botanical Laboratory*, **85**, 191–238.

Churchill, S. P., Griffin, D. & Lewis, M. (1995) Moss diversity of the Tropical Andes. In *Biodiversity and Conservation of Neotropical Forests*, eds. S. P. Churchill, H. Balslev, E. Forero & J. L. Luteyn. New York: New York Botanical Garden, pp. 335–346.

Clark, K. L., Nadkarni, N. M. & Gholz, H. L. (1998) Growth, net production, litter decomposition, and net nitrogen accumulation by epiphytic bryophytes in a tropical montane forest. *Biotropica*, **30**, 12–23.

Cleavitt, N. (2005) Patterns, hypotheses and processes in the biology of rare bryophytes. *Bryologist*, **108**, 554–566.

Cobb, A. R., Nadkarni, N. M., Ramsey, G. A. & Svoboda, A. J. (2001) Recolonization of bigleaf maple branches by epiphytic bryophytes following experimental disturbance. *Canadian Journal of Botany*, **79**, 1–8.

Convey, P. & Lewis-Smith, R. I. (2006) Geothermal bryophyte habitats in the South Sandwich Islands, maritime Antarctic. *Journal of Vegetation Science*, **17**, 529–538.

Crandall-Stotler, B., Stotler, R. E. & Long, D. G. (2009) Morphology and classification of the Marchantiophyta. In *Bryophyte Biology*, 2nd edn, eds. B. Goffinet & A. J. Shaw. Cambridge: Cambridge University Press, pp. 1–54.

Cronberg, N., Molau, U. & Sonesson, M. (1997) Genetic variation in the clonal bryophyte *Hylocomium splendens* at hierarchical geographical scales in Scandinavia. *Heredity*, **78**, 293–301.

Cronberg, N., Natcheva, R. & Hedlund, K. (2006a) Microarthropods mediate sperm transfer in mosses. *Science*, **313**, 1255.

Cronberg, N., Rydgren, K. & Økland, R. H. (2006b) Clonal structure and genet-level sex ratios suggest different roles of vegetative and sexual reproduction in the clonal moss *Hylocomium splendens*. *Ecography*, **29**, 95–103.

Crum, H. A. (1972) The geographic origins of the mosses of North America's eastern deciduous forest. *Journal of the Hattori Botanical Laboratory*, **35**, 269–298.

Crum, H. A. (2001) *Structural Diversity of Bryophytes*. Ann Arbor: University of Michigan.

Crum, H. A. & Anderson, L. E. (1981) *Mosses of Eastern North America*. New York: Columbia University Press.

Davey, M. L. & Currah, R. S. (2006) Interactions between mosses (Bryophyta) and fungi. *Canadian Journal of Botany*, **84**, 1509–1519.

Davison, G. W. H. (1976) Role of birds in moss dispersal. *British Birds*, **69**, 65–66.

De Luca, T. H., Zackrisson, O., Nilsson, M. C. & Sellstedt, A. (2002) Quantifying nitrogen-fixation in feather moss carpets of boreal forests. *Nature*, **419**, 917–920.

DeLucia, E. H., Turnbull, M. H., Walcroft, A. S., *et al.* (2003) The contribution of bryophytes to the carbon exchange for a temperate rainforest. *Global Change Biology*, **9**, 1158–1170.

De Queiroz, A. (2005) The resurrection of oceanic dispersal in historical biogeography. *Trends in Ecology and Evolution*, **20**, 68–73.

Drehwald, U. (2005) Biomonitoring of disturbance in Neotropical rainforests using bryophytes as indicators. *Journal of the Hattori Botanical Laboratory*, **97**, 117–126.

Duckett, J. G. & Ligrone, R. (1992) A survey of diaspore liberation mechanisms and germination patterns in mosses. *Journal of Bryology*, **17**, 335–354.

Duckett, J. G. & Read, D. J. (1991) The use of the fluorescent dye, 3,3′-dihexycarbocyanine iodide, for selective staining of ascomycete fungi associated with liverwort rhizoids and ericoid roots. *New Phytologist*, **118**, 259–272.

Duckett, J. G. & Renzaglia, K. S. (1993) The reproductive biology of the liverwort *Blasia pusilla* L. *Journal of Bryology*, **17**, 541–552.

Duckett, J. G., Prasad, A. K. S. K., Davies, D. A. & Walker, S. (1977) A cytological analysis of the *Nostoc*-bryophyte relationship. *New Phytologist*, **79**, 349–362.

Duckett, J. G., Ligrone, R., Andrews, N. & Renzaglia, K. S. (2000) The enigma of pegged and swollen rhizoids in marchantialean hepatics: a functional explanation. *American Journal of Botany*, **87S**, 6–7.

Duckett, J. G., Burch, J., Fletcher, P. W., *et al.* (2004) In vitro cultivation of bryophytes: a review of practicalities, problems, progress and promise. *Journal of Bryology*, **26**, 3–20.

Duckett, J. G., Russell, J. & Ligrone, R. (2006a) Basidiomycetous endophytes in jungermannialean (leafy) liverworts have novel cytology and species-specific host ranges: a cytological and experimental study. *Canadian Journal of Botany*, **84**, 1075–1093.

Duckett, J. G., Carafa, A. & Ligrone, R. (2006b) A highly differentiated glomeromycotean association with the mucilage-secreting, primitive antipodean liverwort *Treubia* (Treubiaceae): clues to the origins of mycorrhizas. *American Journal of Botany*, **93**, 797–813.

Duckett, J. G., Pressel, S. & Ligrone, R. (2006c) Cornish bryophytes in the Atlantic Arc: cell biology, culturing, conservation and climate change. In *Botanical Links in the Atlantic Arc*, eds. S. J. Leach, C. N. Page, Y. Peytoureau & M. N. Sanford, English Nature and the Botanical Society of the British Isles, Conference Report 24, pp. 165–177.

Duff, R. J., Villareal, J. C., Cargill, D. C. & Renzaglia, K. (2007) Progress and challenges towards developing a phylogeny and classification of hornworts. *Bryologist*, **110**, 214–243.

During, H. J. (1979) Life strategies of bryophytes: a preliminary review. *Lindbergia*, **5**, 2–17.

During, H. J. (1990) The bryophytes of calcareous grasslands. In *Calcareous Grasslands: Ecology and Management*, eds. S. H. Hillier, D. W. H. Walton & D. A. Wells, Bluntisham: Huntingdon, pp. 35–40.

During, H. J. (1992) Ecological classification of bryophytes and lichens. In *Bryophytes and Lichens in Changing Environment*, eds. J. W. Bates & A. M. Farmer. Oxford: Clarendon Press, pp. 1–31.

During, H. J. (1997) Bryophyte diaspore banks. *Advances in Bryology*, **6**, 103–134.

During, H. J. & ter Horst, B. (1983) The diaspore bank of bryophytes and ferns in chalk grasslands. *Lindbergia*, **9**, 57–64.

Edwards, D. (2000) The role of Mid-Paleaozoic mesofossils in the detection of early bryophytes. *Philosophical Transactions of the Royal Society of London B*, **355**, 733–755.

Edwards, D., Duckett, J. G. & Richardson, J. B. (1995) Hepatic characters in the earliest land plants. *Nature*, **374**, 635–636.

Edwards, D., Kerp, D. H. & Hass, H. (1998) Stomata in early land plants: an anatomical and ecophysiological approach. *Journal of Experimental Botany*, **49**, 255–278.

Edwards, D., Axe, L. & Duckett, J. G. (2003) Diversity in conducting cells in early land plants and comparisons with extant bryophytes. *Botanical Journal of the Linnean Society*, **141**, 297–347.

Eldridge, D. J. (1998) Trampling of microphytic crusts on calcareous soils, and its impact on erosion under rain-impacted flow. *Catena*, **33**, 221–239.

Ellis, C. J. & Tallis, J. H. (2000) Climatic control of blanket mire development at Kentra Moss, north-west Scotland. *Journal of Ecology*, **88**, 869–889.

Ellis, C. J. & Tallis, J. H. (2003) Ecology of *Racomitrium lanuginosum* in British blanket mire: evidence from the paleoecological record. *Journal of Bryology*, **25**, 7–15.

Equihua, M. & Usher, M. B. (1993) Impact of carpets of the invasive moss *Campylopus introflexus* on *Calluna vulgaris* regeneration. *Journal of Ecology*, **81**, 359–365.

Eskelinen, A. & Oksanen, J. (2006) Changes in the abundance, composition and species richness of mountain vegetation in relation to summer grazing by reindeer. *Journal of Vegetation Science*, **17**, 245–254.

Fenton, N., Lecomte, N., Légaré, S. & Bergeron, Y. (2005) Paludification in black spruce (*Picea mariana*) forests of eastern Canada: potential factors and management implications. *Forest Ecology and Management*, **213**, 151–159.

Ferguson, N. P. & Lee, J. A. (1980) Some effects of bisulphite and sulphate on the growth of *Sphagnum* species in the field. *Environmental Pollution*, **21**, 59–71.

Ferguson, N. P., Lee, J. A. & Bell, J. N. B. (1978) Effects of sulphur pollutants on the growth of *Sphagnum* species. *Environmental Pollution*, **16**, 151–162.

Floyd, S. K. & Bowman, J. L. (2007) The ancestral developmental tool kit of land plants. *International Journal of Plant Sciences*, **168**, 1–35.

Forrest, L. L., Davis, E. D., Long, D. G., *et al.* (2006) Unravelling the evolutionary history of the liverworts (Marchantiophyta): multiple taxa, genomes, and analyses. *Bryologist*, **109**, 303–334.

Frahm, J. P. (2000) New and interesting mosses from Baltic and Saxon amber. *Lindbergia*, **25**, 33–39.

Frahm, J. P. (2004a) A new contribution to the moss flora of Baltic and Saxon amber. *Review of Paleobotany and Palynology*, **129**, 81–101.

Frahm, J. P. (2004b) Recent developments of commercial products from bryophytes. *Bryologist*, **107**, 277–283.

Frahm, J. P. & Klaus, D. (2001) Bryophytes as indicators of recent climate fluctuations in Central Europe. *Lindbergia*, **26**, 97–104.

Frahm, J. P. & Newton, A. E. (2005) A new contribution to the moss flora of Dominican amber. *Bryologist*, **108**, 526–536.

Frahm, J. P. & Vitt, D. H. (1993) Comparisons between the moss floras of North America and Europe. *Nova Hedwigia*, **56**, 307–333.

Frego, K. A. & Carleton, T. J. (1995a) Microsite conditions and spatial pattern in a boreal bryophyte community. *Canadian Journal of Botany*, **73**, 544–551.

Frego, K. A. & Carleton, T. J. (1995b) Microsite tolerance of four bryophytes in a mature black spruce stand: reciprocal transplants. *Bryologist*, **98**, 452–458.

Frey, W., Hofmann, M. & Hilger, H. H. (2001) The gametophyte–sporophyte junction: unequivocal hints for two evolutionary lines of archegoniate land plants. *Flora*, **196**, 431–445.

Furness, S. B. & Grime, J. P. (1982a) Growth rate and temperature responses in bryophytes. I. An investigation of *Brachythecium rutabulum*. *Journal of Ecology*, **70**, 513–523.

Furness, S. B. & Grime, J. P. (1982b) Growth rate and temperature responses in bryophytes. II. A comparative study of species of contrasted ecology. *Journal of Ecology*, **70**, 525–536.

Gabriel, R. & Bates, J. W. (2003) Responses of photosynthesis to irradiance in bryophytes of the Azores laurel forest. *Journal of Bryology*, **25**, 101–106.

Garbary, D. J. & Renzaglia, K. S. (1998) Bryophyte phylogeny and the evolution of land plants: evidence from development and ultrastructure. In *Bryology for the Twenty-first Century*, eds. J. W. Bates, N. W. Ashton & J. G. Duckett. Leeds: Maney and British Bryological Society, pp. 45–63.

Garbary, D. J., Renzaglia, K. S. & Duckett, J. G. (1993) The phylogeny of land plants: a cladistic analysis based on male gametogenesis. *Plant Systematics and Evolution*, **188**, 237–269.

Garbary, D. J., Miller, A. G., Scrosati, R., Kim, K. Y. & Schofield, W. B. (2008) Distribution and salinity tolerance of intertidal mosses from Nova Scotian salt marshes. *Bryologist*, **111**, 282–291.

Gärdenfors, U. (2000) Population viability analysis in the classification of threatened species: problems and potentials. *Ecological Bulletins*, **48**, 181–190.

Garner, D. B. & Paolillo, D. J., Jr. (1973) On the functioning of stomata in *Funaria*. *Bryologist*, **76**, 423–427.

Gignac, L. D., Nicholson, B. J. & Bayley, S. E. (1998) The utilization of bryophytes in bioclimatic modelling: predicted northward migration of peatlands in the Mackenzie river basin, Canada, as a result of global warming. *Bryologist*, **101**, 572–587.

Gignac, L. D., Halsey, L. A. & Vitt, D. H. (2000) A bioclimatic model for the distribution of *Sphagnum*-dominated peatlands in North America under present climatic conditions. *Journal of Biogeography*, **27**, 1139–1151.

Glime, J. M. (1982) Response of *Fontinalis hypnoides* to seasonal temperature variations. *Journal of the Hattori Botanical Laboratory*, **53**, 181–193.

Glime, J. M. (2007a) *Bryophyte Ecology*. Volume 1. Physiological Ecology. Ebook sponsored by Michigan Technological University and the International Association of Bryologists. Available online at http://www.bryoecol.mtu.edu/.

Glime, J. M. (2007b) Economic and ethnic uses of bryophytes. In *Flora of North America North of Mexico*, Vol. 27. Bryophyta, part 1, ed. Flora of North America Editorial Committee, New York: Oxford University Press, pp. 14–41.

Glime, J. M. & Vitt, D. H. (1984) The physiological adaptations of aquatic musci. *Lindbergia*, **10**, 41–52.

Goebel, K. (1887) [1888]. Morphologische und biologische Studien. I. Über epiphytische Farne und Musciineen. *Annales du Jardin Botanique de Buitenzorg*, **7**, 1–73.

Goffinet, B. (2000) Origin and phylogenetic relationships of bryophytes. In *Bryophyte Biology*, 1st edn, eds. A. J. Shaw & B. Goffinet, Cambridge: Cambridge University Press, pp. 124–149.

Goffinet, B. & Cox, C. J. (2000) Phylogenetic relationships among basal-most arthrodontous mosses with special emphasis on the evolutionary significance of the Funariineae. *Bryologist*, **103**, 212–223.

Goffinet, B., Shaw, A. J. & Cox, C. J. (2004) Phylogenetic inferences in the dung-moss family Splachnaceae from analyses of cpDNA sequence data and implications for the evolution of entomophily. *American Journal of Botany*, **91**, 748–759.

Goffinet, B., Buck, W. R. & Wall, M. A. (2007) *Orthotrichum freyanum* (Orthotrichaceae, Bryophyta), a new epiphytic species from Chile. *Nova Hedwigia, Beiheft*, **131**, 1–11.

Goffinet, B., Buck, W. R. & Shaw, A. J. (2009) Morphology and classification of the Bryophyta. In *Bryophyte Biology*, 2nd edn, eds. B. Goffinet & A. J. Shaw. Cambridge: Cambridge University Press, pp. 55–138.

Goffinet, B. & Shaw, A. J. (eds.) (2009) *Bryophyte Biology*, 2nd edn, Cambridge: Cambridge University Press.

González-Mancebo, J. M., Losada-Lima, A. & McAlister, S. (2003) Host specificity of epiphytic bryophyte communities of a laurel forest on Tenerife (Canary Islands, Spain). *Bryologist*, **106**, 383–394.

González-Mancebo, J. M., Romaguera, F., Losada-Lima, A. & Suarez, A. (2004) Epiphytic bryophytes growing on *Laurus azorica* (Seub.) Franco in three laurel forest areas in Tenerife (Canary Islands). *Acta Oecologica*, **25**, 159–167.

Grace, M. (1995) A key to the growth forms of mosses and liverworts and guide to their educational value. *Journal of Biological Education*, **29**, 272–278.

Gradstein, S. R. (1990) Morphology and classification of the Hepaticae: an introduction. In *Chemistry and Chemical Taxonomy of Bryophytes*, eds. H. D. Zinsmeister & R. Mues. Oxford: Oxford University Press, pp. 3–17.

Gradstein, S. R. (1992a) The vanishing Tropical Rain Forest as an environment for bryophytes and lichens. In *Bryophytes and Lichens in a Changing Environment*, eds. J. W. Bates & A. M. Farmer. Oxford: Oxford University Press, pp. 232–256.

Gradstein, S. R. (1992b) Threatened bryophytes of the neotropical rain forest: a status report. *Tropical Bryology*, **6**, 83–93.

Gradstein, S. R. (1993) New fossil hepaticae preserved in amber of the Dominican Republic. *Nova Hedwigia*, **57**, 353–374.

Gradstein, S. R. (1995) Diversity of Hepaticae and Anthocerotae in montane forests of the tropical Andes. In *Biodiversity and Conservation of Neotropical Montane Forests*, eds. S. P. Churchill, H. Balslev, E. Forero & J. L. Luteyn. New York: New York Botanical Garden, pp. 321–334.

Gradstein, S. R. & Weber, W. A. (1982) Bryogeography of the Galapagos Islands. *Journal of the Hattori Botanical Laboratory*, **52**, 127–152.

Gradstein, S. R. & Wilson, R. (2008) Protonemal neoteny in bryophytes. In *Bryology in the New Millenium*, eds. H. Mohamed, B. H. Bakar, A. N. Boyce & P. N. Lee. Kuala Lumpur, pp. 1–12.

Gradstein, S. R., Klein, R., Kraut, L., *et al.* (1992) Phytochemical and morphological support for the existence of two species in *Monoclea* (Hepaticae). *Plant Systematics and Evolution*, **180**, 115–135.

Gradstein, S. R., Churchill, S. P. & Salazar-Allen, N. (2001) *Guide to the Bryophytes of Tropical America*. New York: New York Botanical Garden.

Gradstein, S. R., Reiner-Drehwald, M. E. & Muth, H. (2003) Über die Identität der neuen Aquarienpflanze *Pellia endiviifolia*. *Aqua Planta*, **3**, 88–95.

Graham, L. E. & Wilcox, L. W. (2000) The origin of alternation of generations in land plants: a focus on matrotrophy and hexose transport. *Philosophical Transactions of the Royal Society of London B*, **355**, 757–767.

Graham, L. E., Cook, M. E. & Busse, J. S. (2000) The origin of plants: Body plan changes contributing to a major evolutionary radiation. *Proceedings of the National Academy of Sciences of the USA*, **97**, 4535–4540.

Graham, L. E., Wilcox, L. W. Cook, M. E. & Gensel, P. G. (2004) Resistant tissues of modern marchantioid liverworts resemble enigmatic Early Paleaozoic microfossils. *Proceedings of the National Academy of Sciences of the USA*, **101**, 11025–11029.

Graham, L. E., Kodner, R. B., Fisher, M. M., *et al.* (2006) Early land plant adaptations to terrestrial stress: a focus on phenolics. In *The Evolution of Plant Physiology*, eds. A. R. Hemsely & I. Poole. Amsterdam: Elsevier, pp. 155–169.

Grau, O., Grytnes, J.-A. & Birks, H. J. B. (2007) A comparison of altitudinal species richness of bryophytes with other plant groups in Nepal, Central Himalaya. *Journal of Biogeography*, **34**, 1907–1915.

Groeneveld, E. V. G. & Rochefort, L. (2005) *Polytrichum strictum* as a solution to frost heaving in disturbed ecosystems: a case study with milled peatlands. *Restoration Ecology*, **13**, 74–82.

Grundmann, M., Ansell, S. W., Russell, S. J., Koch, M. A. & Vogel, J. C. (2007) Genetic structure of the widespread and common Mediterranean bryophyte *Pleurochaete squarrosa* (Brid.) Lindb. (Pottiaceae): evidence from nuclear and plastidic DNA sequence variation and allozymes. *Molecular Ecology*, **16**, 709–722.

Gunnarsson, U. & Söderström, L. (2007) Can artificial introduction of diaspore fragments work as a conservation tool for maintaining populations of the rare peatmoss *Sphagnum angermanicum*? *Biological Conservation*, **135**, 450–458.

Gunnarsson, U., Rydin, H. & Sjörs, H. (2000) Diversity and pH changes after 50 years on the boreal mire Skattlösbergs Stormosse, Central Sweden. *Journal of Vegetation Science*, **11**, 277–286.

Gunnarsson, U., Hassel, K. & Söderström, L. (2005) Genetic structure of the endangered peatmoss *Sphagnum angermanicum* in Sweden: a result of historic or contemporary processes? *Bryologist*, **108**, 194–202.

Hallingbäck, T. (2003) Including bryophytes in international conventions: a success story from Europe. *Journal of the Hattori Botanical Laboratory*, **9**, 201–214.

Hallingbäck, T. & Hodgetts, N. (2000) *Mosses, Liverworts, and Hornworts. Status Survey and Conservation Action Plan for Bryophytes*. Gland: IUCN.

Hanski, I. (1999) *Metapopulation Ecology*. Oxford: Oxford University Press.

Harmens, H., Buse, A., Büker, P., *et al.* (2004) Heavy metal concentrations in European mosses: 2000/2001 survey. *Journal of Atmospheric Chemistry*, **49**, 425–436.

Harvey-Gibson, R. J. & Miller-Brown, D. (1927) Fertilization of bryophyta. *Annals of Botany*, **41**, 190–191.

Hasegawa, J. (1988) A proposal for a new system of the Anthocerotae, with a revision of the genera. *Journal of the Hattori Botanical Laboratory*, **64**, 87–95.

Hassel, K. & Söderström, L. (2005) The expansion of the alien mosses *Orthodontium lineare* and *Campylopus introflexus* in Britain and continental Europe. *Journal of the Hattori Botanical Laboratory*, **97**, 183–193.

Hassel, K., Pedersen, B. & Söderström, L. (2005) Changes in life-history traits in an expanding moss species: phenotypic plasticity or genetic differentiation? A reciprocal transplantation experiments with *Pogonatum dentatum*. *Ecography*, **28**, 71–80.

Hébant, C. (1977) The conducting tissues of bryophytes. *Bryophytorum Bibliotheca*, **10**, 1–157.

Heber, U., Bukhov, N. G., Shuvalov, V., Kobayashi, Y. & Loange, O. L. (2001) Protection of the photosynthetic apparatus against damage by excessive illumination in homoiohydric leaves and poikylohydric mosses and lichens. *Journal of Experimental Botany*, **52**, 1999–2006.

Heckman, D. S., Geiser, D. M., Eidell, B. R., *et al.* (2001) Molecular evidence for the early colonization of land by fungi and plants. *Science*, **293**, 1129–1133.

Hedenäs, L. (2001) The importance of phylogeny and habitat factors in explaining gametophytic character states in European Amblystegiaceae. *Journal of Bryology*, **23**, 205–219.

Hedderson, T. A. & Nowell, T. L. (2006) Phylogeography of *Homalothecium sericeum* (Hedw.) Br. Eur.: toward a reconstruction of glacial survival and postglacial migration. *Journal of Bryology*, **28**, 283–292.

Hedges, S. B., Blair, J. E., Venturi, M. L. & Shoe, J. L. (2004) A molecular timescale of eukaryote evolution and the rise of complex multicellular life. *BMC Evolutionary Biology*, **4**, article 2.

Heegaard, E. (1997) Ecology of *Andreaea* in western Norway. *Journal of Bryology*, **19**, 527–636.

Heilmann-Clausen, J., Aude, E. & Christensen, M. (2005) Cryptogam communities on decaying deciduous wood: does tree species diversity matter? *Biodiversity and Conservation*, **14**, 2061–2078.

Heinken, T., Lees, R., Raudnitschka, D. & Runge, S. (2001) Epizoochorous dispersal of bryophyte stem fragments by roe deer (*Capreolus capreolus*) and wild boar (*Sus scrofa*). *Journal of Bryology*, **23**, 293–300.

Heinrichs, J., Gradstein, S. R., Wilson, R. & Schneider, H. (2005) Towards a natural classification of liverworts (Marchantiophyta) based on the chloroplast gene *rbc*L. *Cryptogamie. Bryologie*, **26**, 131–150.

Heinrichs, J., Hentschel, J., Wilson, R., Feldberg, K. & Schneider, H. (2007) Evolution of leafy liverworts (Jungermannidae, Marchantiophyta): estimating divergence times from chloroplast DNA sequences using penalized likelihood with integrated fossil evidence. *Taxon*, **56**, 31–44.

He-Nygrén, X., Juslén, A., Ahonen, I., Glenny, D. & Piippo, S. (2006) Illuminating the evolutionary history of liverworts (Marchantiophyta): towards a natural classification. *Cladistics*, **22**, 1–31.

Heywood, V. H. & Iriondo, J. M. (2003) Plant conservation: old problems, new perspectives. *Biological Conservation*, **113**, 321–335.

Higuchi, S., Kawamura, M., Miyajima, I., *et al.* (2003) Morphology and phylogenetic position of a mat-forming green plant from acidic rivers in Japan. *Journal of Plant Research*, **116**, 461–467.

Hock, Z., Szövényi, P. & Toth, Z. (2004) Seasonal variation in the bryophyte diaspore bank of open grasslands on dolomite rocks. *Journal of Bryology*, **26**, 285–292.

Hock, Z., Szövényi, P., Schneller, J. J., Toth, Z. & Urmi, E. (2008) Bryophyte diaspore bank: a genetic memory? Genetic structure and genetic diversity of surface populations and diaspore bank in the liverwort *Mannia fragrans* (Aytoniaceae). *American Journal of Botany*, **95**, 542–548.

Hoekstra, F. A. (2005) Differential longevities in desiccated anhydrobiotic plant systems. *Integrative and Comparative Biology*, **45**, 725–733.

Hogg, P., Squires, P. & Fitter, A. H. (1995) Acidification, nitrogen deposition and rapid vegetational change in a small valley mire in Yorkshire. *Biological Conservation*, **71**, 143–153.

Holz, I. & Gradstein, S. R. (2005) Cryptogamic epiphytes in primary and recovering upper montane oak forests of Costa Rica: species richness, community composition and ecology. *Plant Ecology*, **178**, 547–560.

Holz, I., Gradstein, R., Heinrichs, J. & Kappelle, M. (2002) Bryophyte diversity, microhabitat differentiation, and distribution of life forms in Costa Rican upper montane Quercus forest. *Bryologist*, **105**, 334–348.

Hornschuh, M., Grotha, R. & Kutschera, U. (2002) Epiphytic bacteria associated with the bryophyte *Funaria hygrometrica*: effects of Methylobacterium strains on protonema development. *Plant Biology*, **4**, 682–687.

Humphries, C. J. & Parenti, L. R. (1999) *Cladistic Biogeography*, 2nd edn. Oxford: Oxford University Press.

Hutchinson, T. C. & Scott, M. (1988) The response of the feather moss *Pleurozium schreberi* (Brid.) Mitt. to five years of simulated acid precipitation in the Canadian boreal forest. *Canadian Journal of Botany*, **66**, 82–88.

Hutsemékers, V., Dopagne, C. & Vanderpoorten, A. (2008) How far and how fast do bryophytes disperse at the landscape scale? *Diversity and Distributions*, **14**, 483–492.

Huttunen, S., Hedenäs, L., Ignatov, M. S., Devos, N. & Vanderpoorten, A. (2008) Origin and evolution of the northern hemisphere disjunction in the moss genus *Homalothecium* (Brachytheciaceae). *American Journal of Botany*, **95**, 720–730.

Hylander, K. & Jonsson, B. G. (2007) The conservation ecology of cryptogams. *Biological Conservation*, **135**, 311–314.

Hyvönen, J. & Piippo, S. (1993) Cladistic analysis of the hornworts (Anthocerotophyta). *Journal of the Hattori Botanical Laboratory*, **74**, 105–119.

Hyvönen, J., Koponen, T. & Norris, D. H. (1987) Human influence on the moss flora of tropical rain forest in Papua New Guinea. *Symposia Biologia Hungarica*, **35**, 621–629.

Ignatov, M. S. (1992) *Bryokhutuliina jurassica*, gen. et spec. nova, a remarkable fossil moss from Mongolia. *Journal of the Hattori Botanical Laboratory*, **71**, 377–388.

Ihl, C. & Barboza, P. S. (2007) Nutritional value of moss for Arctic ruminants: a test with muskoxen. *Journal of Wildlife Management*, **71**, 752–758.

Ines-Sastre, J. & Tan, B. C. (1995) Problems of bryophyte conservation in the tropics: a discussion, with case example from Puerto Rico and the Philippines. *Carribean Journal of Science*, **31**, 200–206.

Ireland, R. R. (1971) Moss pseudoparaphyllia. *Bryologist*, **74**, 313–330.

IUCN (2001) iucn.org/themes/ssc/redlists/RLcats2001booklet.html

Johnson, D. S. (1904) The development and relationship of *Monoclea*. *Botanical Gazette*, **38**, 185–205.

Jongmans, A. G., van Breemen, N., Gradstein, S. R. & van Oort, F. (2001) How liverworts build hanging gardens from volcanic ash in Costa Rica. *Catena*, **44**, 13–22.

Jonsgard, B. & Birks, H. H. (1995) Late-glacial mosses and environmental reconstructions at Kråkenes, western Norway. *Lindbergia*, **20**, 64–82.

Jonsson, B. G. (1993) The bryophyte diaspore bank and its role after small-scale disturbance in a boreal forest. *Journal of Vegetation Science*, **4**, 819–826.

Jordan, C. F. (1995) *Conservation*. New York: Wiley.

Kelch, D. G., Driskell, A. & Mishler, B. D. (2004) Inferring phylogeny using genomic characters: a case study using land plant plastomes. *Monographs in Systematic Botany from the Missouri Botanical Garden*, **98**, 3–12.

Kenrick, P. & Crane, P. R. (1997a) *The Origin and Early Diversification of Land Plants. A Cladistic Study*. Washington, DC: Smithsonian Institution Press.

Kenrick, P. & Crane, P. R. (1997b) The origin and early evolution of plants on land. *Nature*, **389**, 33–39.

Ketner-Oostra, R. & Sykora, K. V. (2004) Decline of lichen-diversity in calcium-poor coastal dune vegetation since the 1970s, related to grass and moss encroachment. *Phytocoenologia*, **34**, 521–549.

Kimmerer, R. W. (1993) Disturbance and dominance in *Tetraphis pellucida*: a model of disturbance frequency and reproductive mode. *Bryologist*, **96**, 73–79.

Kimmerer, R. W. (1994) Ecological consequences of sexual versus asexual reproduction in *Dicranum flagellare* and *Tetraphis pellucida*. *Bryologist*, **97**, 20–25.

Klanderud, K. & Totland, O. (2005) Simulated climate change altered dominance hierarchies and diversity of an alpine biodiversity hotspot. *Ecology*, **86**, 2047–2054.

Kodner, R. B. & Graham, L. E. (2001) High-temperature, acid-hydrolyzed remains of *Polytrichum* (Musci, Polytrichaceae) resemble enigmatic Silurian-Devonian tubular microfossils. *American Journal of Botany*, **88**, 462–466.

Konopka, A. S., Herendeen, P. S., Smith-Merrill, G. L. & Crane, P. R. (1997) Sporophytes and gametophytes of Polytrichaceae from the late Campanian (Late Cretaceous) of Georgia, USA. *International Journal of Plant Sciences*, **158**, 489–499.

Konopka, A. S., Herendeen, P. S. & Crane, P. R. (1998) Sporophytes and gametophytes of Dicranaceae from the Santonian (Late Cretaceous) of Georgia, USA. *American Journal of Botany*, **85**, 714–723.

Kooijman, A. M. (1993) Causes of the replacement of *Scorpidium scorpioides* by *Calliergonella cuspidata* in eutrophicated rich fens. 1. Field studies. *Lindbergia*, **18**, 78–84.

Kooijman, A. M. & Bakker, C. (1993) Causes of the replacement of *Scorpidium scorpioides* by *Calliergonella cuspidata* in eutrophicated rich fens: 2. Experimental studies. *Lindbergia*, **18**, 123–130.

Koranda, M., Kerschbaum, S., Wanek, W., Zechmeister, H. & Richter, A. (2007) Physiological responses of bryophytes *Thuidium tamariscinum* and *Hylocomium splendens* to increased nitrogen deposition. *Annals of Botany*, **99**, 161–169.

Korpelainen, H., Pohjamo, M. & Laaka-Lindberg, S. (2005) How efficiently does bryophyte dispersal lead to gene flow? *Journal of the Hattori Botanical Laboratory*, **97**, 195–205.

Kottke, I., Beiter, A., Weiss, M., *et al.* (2003) Heterobasidiomycetes form symbiotic associations with hepatics: Jungermanniales have sebacinoid mycobionts while *Aneura pinguis* (Metzgeriales) is associated with a *Tulasnella* species. *Mycological Research*, **107**, 957–968.

Krings, M., Taylor, T. N., Hass, H., *et al.* (2007) Fungal endophytes in a 400-million-yr-old land plant: infection pathways, spatial distribution and host responses. *New Phytologist*, **174**, 648–657.

Krommer, V., Zechmeister, H. G., Roder, I., Scharf, S. & Hanus-Illnar, A. (2007) Monitoring atmospheric pollutants in the biosphere reserve Wienerwald by a combined approach of biomonitoring methods and technical measurements. *Chemosphere*, **67**, 1956–1966.

Kruijer, H. (2002) Hypopterygiaceae of the world. *Blumea*, suppl. **13**, 1–388.

Kürschner, H. (2003) The *Riccietum jovet-astii–argenteolimbatae* ass. nov. of the Jabal Arays area, Yemen: life strategies of a remarkable xerotropical African bryophyte community. *Nova Hedwigia*, **76**, 399–413.

Kürschner, H. & Parolly, G. (1999) The *Epipterygio-Riccietum frostii* ass. nov.: ecology and life strategies of an ephemeral bryophyte community in western Turkey. *Lindbergia*, **24**, 84–92.

Kürschner, H. & Parolly, G. (2004) Phytomass and water-storing capacity of epiphytic rain forest communities in S Ecuador. Ecosociological studies in Ecuadorian bryophyte communities. IV. *Botanische Jahrbücher für Systematik*, **125**, 489–504.

Kürschner, H. & Parolly, G. (2005) Ecosociological studies in Ecuadorian bryophyte communities. III. Life forms, life strategies, and ecomorphology of the submontane and montane epiphytic vegetation of S Ecuador. *Nova Hedwigia*, **80**, 89–114.

Laaka-Lindberg, S. (2000) Substrate preference and reproduction in *Lophozia silvicola* (Hepaticopsida) in southern Finland. *Annales Botanici Fennici*, **37**, 85–93.

Laaka-Lindberg, S. (2005) Reproductive phenology in the leafy hepatic *Lophozia silvicola* Buch in southern Finland. *Journal of Bryology*, **27**, 253–259.

Laaka-Lindberg, S., Korpelainen, H. & Pohjamo, M. (2003) Dispersal of asexual propagules in bryophytes. *Journal of Hattori Botanical Laboratory*, **93**, 319–330.

La Farge-England, C. (1996) Growth form, branching pattern, and perichaetial position in mosses: cladocarpy and pleurocarpy re-defined. *Bryologist*, **99**, 170–186.

Lambeck, R. J. (1997) Focal species: a multi-species umbrella for nature conservation. *Conservation Biology*, **11**, 849–856.

Laurance, W. F. (1998) A crisis of the making: responses of Amazonian forests to land use and climate change. *Trends in Ecology and Evolution*, **13**, 411–412.

Lavoie, C. & Rochefort, L. (1996) The natural revegetation of a harvested peatland in southern Québec: a spatial and dendroecological analysis. *Ecoscience*, **3**, 101–111.

Lavoie, M., Paré, D., Fenton, N., Groot, A. & Taylor, K. (2005) Paludification and management of forested peatlands in Canada: a literature review. *Environmental Reviews*, **13**, 21–50.

LeBlanc, F. & De Sloover, J. (1970) Relation between industrialization and the distribution and growth of epiphytic lichens and mosses in Montreal. *Canadian Journal of Botany*, **48**, 1485–1496.

Lee, J. A. (1998) Unintentional experiments with terrestrial ecosystems: ecological effects of sulphur and nitrogen pollutants. *Journal of Ecology*, **86**, 1–12.

Leitgeb, H. (1876) Über verzweigte Moossporogonien. *Mitteilungen des Naturwissenschaftlichen Vereines für Steiermark*, **13**, 1–20.

Léon-Vargas, Y., Engwald, S. & Proctor, M. C. F. (2006) Microclimate, light adaptation and desiccation tolerance of epiphytic bryophytes in two Venezuelan cloud forests. *Journal of Biogeography*, **33**, 901–913.

Levitt, J. (1980) *Responses of Plants to Environmental Stresses*, 2nd edn. New York: Academic Press.

Lewis Smith, R. I. (1999) Biological and environmental characteristics of three cosmopolitan mosses dominant in continental Antarctica. *Journal of Vegetation Science*, **10**, 231–242.

Ligrone, R. & Duckett, J. G. (1994) Cytoplasmic polarity and endoplasmic microtubules associated with the nucleus and organelles are ubiquitous features of food-conducting cells in bryoid mosses (Bryophyta). *New Phytologist*, **127**, 601–614.

Ligrone, R., Duckett, J. G. & Renzaglia, K. S. (1993) The gametophyte–sporophyte junction in land plants. *Advances in Botanical Research*, **19**, 231–317.

Ligrone, R., Duckett, J. G. & Renzaglia, K. S. (2000) Conducting tissues and phyletic relationships of bryophytes. *Philosophical Transactions of the Royal Society*, **355**, 795–813.

Ligrone, R., Carafa, A., Lumin, E., *et al.* (2007) Glomeromycotean associations in liverworts: a molecular, cellular, and taxonomic analysis. *American Journal of Botany*, **94**, 1756–1777.

Ligrone, R., Carafa, A., Duckett, J. G., Renzaglia, K. S. & Ruel, K. (2008) Immunocytochemical detection of lignin-related epitopes in cell walls in bryophytes and the charalean green alga *Nitella*. *Plant Systematics and Evolution*, **270**, 257–272.

Limpens, J., Tomassen, H. B. M. & Berendse, F. (2003) Expansion of *Sphagnum fallax* in bogs: striking the balance between N and P availability. *Journal of Bryology*, **25**, 83–90.

Lindenmayer, D. B., Margules, C. R. & Botkin, D. (2000) Indicators of forest sustainability biodiversity: the selection of forest indicator species. *Conservation Biology*, **14**, 941–950.

Lindenmayer, D. B., Manning, A. D., Smith, P. L., *et al.* (2002) The focal-species approach and landscape restoration: a critique. *Conservation Biology*, **16**, 338–345.

Longton, R. E. (1992) The role of bryophytes and lichens in terrestrial ecosystems. In *Bryophytes in a Changing Environment*, eds. J. W. Bates & A. M. Farmer. Oxford: Clarendon Press, pp. 32–76.

Longton, R. E. (1997) Reproductive biology and life-history strategies. *Advances in Bryology*, **6**, 65–101.

Longton, R. E. & Hedderson, T. A. (2000) What are rare species and why conserve them? *Lindbergia*, **25**, 53–61.

Longton, R. E. & Schuster, R. M. (1983) Reproductive biology. In *New Manual of Bryology*, ed. R. M. Schuster. Nichinan: Hattori Botanical Laboratory, pp. 386–462.

Lopez, J., Retuerto, R. & Carballeira, A. (1997) D665/D665a index vs frequencies as indicators of bryophyte response to physicochemical gradients. *Ecology*, **78**, 261–271.

Lucas, J. R. & Renzaglia, K. S. (2002) Structure and function of hornwort stomata. *Proceedings Microscopy and Microanalysis*, **8**, 1090–1091.

Malcolm, B. & Malcolm, N. (2006) *Mosses and Other Bryophytes: An Illustrated Glossary*. Nelson: Micro-Optics Press.

Manzke, W. (2004) Zur Verbreitung und Bestandssituation von *Notothylas orbicularis, Anthoceros neesii, Anthoceros agrestis, Phaeoceros carolinianus* und *Riccia ciliata* auf Stoppelfeldern im Vogelsberg (Hessen). *Hessische Floristische Briefe*, **53**, 53–65.

Marino, P. C., Raguso, R. A. & Goffinet, B. (2009) Insect mediated spore dispersal in the moss family Splachnaceae: odour chemistry, ecology and evolution. *Symbiosis*, **47**, 61–76.

Markert, B. A., Breure, A. M. & Zechmeister, H. G. (2003) *Bioindicators and Biomonitors. Principles, Concepts and Applications*. Oxford: Elsevier.

Marrero-Gomez, M. V., Banares-Baudet, A. & Carque-Alamo, E. (2003) Plant resource conservation planning in protected natural areas: an example from the Canary Islands, Spain. *Biological Conservation*, **113**, 399–410.

Marschall, M. & Proctor, M. C. F. (2004) Are bryophytes shade plants? Photosynthetic light responses and proportions of chlorophyll a, chlorophyll b and total carotenoids. *Annals of Botany*, **94**, 593–603.

Martinez-Abaigar, J., Nunez-Olivera, E. & Beaucourt, N. (2002) Short-term physiological responses of the aquatic liverwort *Jungermannia exsertifolia* subsp. *cordifolia* to KH_2PO_4 and anoxia. *Bryologist*, **105**, 86–95.

McArthur, R. H. & Wilson, E. O. (1967) *The Theory of Island Biogeography*. Princeton: Princeton University Press.

McCourt, R. M., Delwiche, C. F. & Karol, K. G. (2004) Charophyte algae and land plant origins. *Trends in Ecology and Evolution*, **19**, 661–666.

McCune, B., Amsberry, K. A., Camacho, F. J., *et al.* (1997) Vertical profile of epiphytes in a Pacific Northwest old-growth forest. *Northwest Science*, **71**, 145–152.

McDaniel, S. F. & Shaw, A. J. (2003) Phylogeographic structure and cryptic speciation in the trans-Antarctic moss *Pyrrhobryum mnioides*. *Evolution*, **57**, 205–215.

McDowall, R. M. (2004) What biogeography is: a place for process. *Journal of Biogeography*, **31**, 344–351.

McGlone, M. S. (2005) Goodbye Gondwana. *Journal of Biogeography*, **32**, 739–740.

McIntosh, T. & Miles, W. (2005) Comments on rare and interesting bryophytes in garry oak ecosystems, British Columbia, Canada. *Journal of the Hattori Botanical Laboratory*, **97**, 263–269.

McLetchie, D. N. & Puterbaugh, M. N. (2000) Population sex ratios, sex-specific clonal traits and tradeoffs among these traits in the liverwort *Marchantia inflexa*. *Oikos*, **90**, 227–237.

McLetchie, D. N. & Stark, L. R. (2006) Sporophyte and gametophyte generations differ in their thermotolerance response in the moss *Microbryum*. *Annals of Botany*, **97**, 505–511.

Menand, B., Yi, K., Hoffann, L., *et al.* (2007) An ancient mechanism controls the development of cells with a rooting function in land plants. *Science*, **316**, 1477–1480.

Merrifield, K. & Ingham, R. R. (1998) Nematodes and other aquatic invertebrates in *Eurhynchium oreganum* from Mary's Peak, Oregon Coast Range. *Bryologist*, **101**, 505–511.

Meyer, H. & Santarius, K. A. (1998) Short-term thermal acclimatation and heat tolerance of gametophytes of mosses. *Oecologia*, **115**, 1–8.

Miles, C. J. & Longton, R. E. (1992) Deposition of moss spores in relation to distance from parent gametophytes. *Journal of Bryology*, **17**, 355–368.

Miller, N. G. & McDaniel, S. F. (2004) Bryophyte dispersal inferred from colonization of an introduced substratum on Whiteface Mountain, New York. *American Journal of Botany*, **91**, 1173–1182.

Miller, N. G. & Trigoboff, N. (2001) A European feather moss, *Pseudoscleropodium purum*, naturalized widely in New York State in cemeteries. *Bryologist*, **104**, 98–103.

Minami, A., Nagao, M., Arakawa, K., Fujikawa, S. & Takezawa, D. (2006) Physiological and morphological adaptations associated with the development of freezing tolerance in the moss *Physcomitrella patens*. In *Cold Hardiness in Plants: Molecular Genetics, Cell Biology and Physiology*, eds. T. H. H. Chen, M. Uemura & S. Fujikawa. Wallingford: CAB International, pp. 138–152.

Mishler, B. D. & Churchill, S. P. (1984) A cladistic approach to the phylogeny of the 'bryophytes'. *Brittonia*, **36**, 406–424.

Mishler, B. D. & Churchill, S. P. (1985) Transition to a land flora: phylogenetic relationships of the green algae and bryophytes. *Cladistics*, **1**, 305–328.

Mishler, B. D., Lewis, L. A., Buchheim, M. A., *et al.* (1994) Phylogenetic relationships of the 'green algae' and 'bryophytes.' *Annals of the Missouri Botanical Garden*, **81**, 451–483.

Mittelbach, G. G., Schemske, D. W., Cornell, H. V., *et al.* (2007) Evolution and the latitudinal diversity gradient: speciation, extinction and biogeography. *Ecology Letters*, **10**, 315–331.

Moe, B. & Botnen, A. (2000) Epiphytic vegetation on pollarded trunks of *Fraxinus excelsior* in four different habitats at Grinde, Leikanger, western Norway. *Plant Ecology*, **151**, 143–159.

Molau, U. & Alatalo, J. M. (1998) Responses of subarctic-alpine plant communities to simulated environmental change: biodiversity of bryophytes, lichens and vascular plants. *Ambio*, **27**, 322–329.

Morgan, J. W. (2006) Bryophyte mats inhibit germination of non-native species in burnt temperate native grassland remnants. *Biological Invasions*, **8**, 159–168.

Mouvet, C. & Clavieri, B. (1999) Localization of copper accumulated in *Rhynchostegium riparioides* using sequential chemical extraction. *Aquatic Botany*, **63**, 1–10.

Mouvet, C., Morhain, E., Sutter, C. & Couturieux, N. (1993) Aquatic mosses for the detection and follow-up of accidental discharges in surface waters. *Water, Air and Soil Pollution*, **66**, 333–348.

Muir, P. S., Norman, K. N. & Sikes, K. G. (2006) Quantity and value of commercial moss harvest from forests of the Pacific Northwest and Appalachian regions of the US. *Bryologist*, **109**, 197–214.

Müller, P., Li, X. P. & Niyogi, K. K. (2001) Non-photochemical quenching. A response to excess light energy. *Plant Physiology*, **125**, 1558–1566.

Muñoz, J., Felicisimo, A. M., Cabezas, F., Burgaz, A. R. & Martinez, I. (2004) Wind as a long-distance dispersal vehicle in the southern Hemisphere. *Science*, **304**, 1144–1147.

Nadkarni, N. M. (1984) The biomass and nutrient capital of epiphytes in a neotropical cloud forest, Monteverde. *Biotropica*, **15**, 1–9.

Newton, A. E. & Mishler, B. D. (1994) The evolutionary significance of asexual reproduction in mosses. *Journal of the Hattori Botanical Laboratory*, **76**, 127–145.

Newton, A. E., Cox, C. J., Duckett, J. G., *et al.* (2000) Evolution of the major moss lineages: Phylogenetic analyses based on multiple gene sequences and morphology. *Bryologist*, **103**, 187–211.

Newton, A. E., Wikström, N., Bell, N., Forrest, L. L. & Ignatov, M. S. (2007) Dating the diversification of the pleurocarpous mosses. In *Pleurocarpous Mosses. Systematics and Evolution*, eds. A. E. Newton & R. S. Tangney. Boca Raton: Taylor & Francis, pp. 337–366.

Nishiyama, T., Fujita, T., Shin-I, T., *et al.* (2003) Comparative genomics of *Physcomitrella patens* gametophytic transcriptome and *Arabidopsis thaliana*: implication for land plant evolution. *Proceedings of the National Academy of Sciences of the USA*, **100**, 8007–8012.

Norden, B., Paltto, H., Götmark, F. & Wallin, K. (2007) Indicators of biodiversity, what do they indicate? Lessons for conservation of cryptogams in oak-rich forest. *Biological Conservation*, **135**, 369–379.

Økland, R. H. (1994) Patterns of bryophyte associations at different scales in a Norwegian boreal spruce forest. *Journal of Vegetation Science*, **5**, 127–138.

Økland, R. H. (2000) Population biology of the clonal moss *Hylocomium splendens* in Norwegian boreal spruce forests. 5. Vertical dynamics of individual shoot segments. *Oikos*, **88**, 449–469.

Økland, R. H. & Bakkestuen, V. (2004) Fine-scale spatial patterns in populations of the clonal moss *Hylocomium splendens* partly reflect structuring processes in the boreal forest floor. *Oikos*, **106**, 565–575.

Økland, R. H. & Økland, T. (1996) Population biology of the clonal moss *Hylocomium splendens* in Norwegian boreal spruce forests. II. Effects of density. *Journal of Ecology*, **84**, 63–69.

Oliver, M. J. (1991) Influence of protoplasmic water loss on the control of protein synthesis in the desiccation-tolerant moss *Tortula ruralis*: ramifications for a repair-based mechanism of desiccation tolerance. *Plant Physiology*, **97**, 1501–1511.

Oliver, M. J., Tuba, Z. & Mishler, B. D. (2000) The evolution of vegetative desiccation tolerance in land plants. *Plant Ecology*, **151**, 85–100.

Oliver, M. J., Velten, J. & Mishler, B. D. (2005) Desiccation tolerance in bryophytes: a reflection of the primitive stategy for plant survival in dehydrating habitats? *Integrative and Comparative Biology*, **45**, 788–799.

O'Neill, K. P. (2000) Role of bryophyte-dominated ecosystems in the global carbon budget. In *Bryophyte Biology*, 1st edn, eds. A. J. Shaw & B. Goffinet. Cambridge: Cambridge University Press, pp. 344–368.

O'Shea, B. J. (1997) The mosses of sub-Saharan Africa 2. Endemism and biodiversity. *Tropical Bryology*, **13**, 75–85.

Oyesiku, O. O. & Ogunkolade, O. R. (2006) The relationship between the Nigerian garden snail *Limicolaria aurora* and the moss *Hyophila crenulata*. *Journal of Bryology*, **28**, 104–107.

Parenti, L. R. (1980) A phylogenetic analysis of the land plants. *Biological Journal of the Linnean Society*, **13**, 225–242.

Parker, J. D., Burkepile, D. E., Collins, D. O., Kubanek, J. & Hay, M. E. (2000) Stream mosses as chemically-defended refugia for freshwater macroinvertebrates. *Oikos*, **116**, 302–312.

Parsons, G., Cairns, A., Johnson, C. N., *et al.* (2007) Bryophyte dispersal by flying foxes: a novel discovery. *Oecologia*, **152**, 112–114.

Pärtel, M., Helm, A., Ingerpuu, N., Reier, Ü. & Tuvi, E.-V. (2004) Conservation of Northern European plant diversity: the correspondence with soil pH. *Biological Conservation*, **120**, 525–531.

Pearce, I. S. K. & van der Wal, R. (2002) Effects of nitrogen deposition on growth and survival of montane *Racomitrium lanuginosum* heath. *Biological Conservation*, **104**, 83–89.

Peck, J. E. (2006) Towards sustainable commercial moss harvest in the Pacific Northwest of North America. *Biological Conservation*, **128**, 289–297.

Peck, J. E. & Christy, J. A. (2006) Putting the stewardship concept into practice: commercial moss harvest in northwestern Oregon, USA. *Forest Ecology and Management*, **225**, 225–233.

Peck, J. E. & McCune, B. C. (1998) Commercial moss harvest in northwestern Oregon: biomass and accumulation of epiphytes. *Biological Conservation*, **86**, 299–305.

Pentecost, A. (1998) Some observations on the biomass and distribution of cryptogamic epiphytes in the upper montane forest of Rwenzori mountains, Uganda. *Global Ecology and Biogeography Letters*, **7**, 273–284.

Pharo, E. J. & Blanks, P. A. M. (2000) Managing a neglected component of biodiversity: a study of bryophyte diversity in production forests of Tasmania's northeast. *Australian Forestry*, **63**, 128–135.

Pharo, E. J., Beattie, A. J. & Pressey, R. L. (2000) Effectiveness of using vascular plants to select reserves for bryophytes and lichens. *Biological Conservation*, **96**, 371–378.

Pharo, E. J., Kirkpatrick, J. B., Gilfedder, L., Mendel, L. & Turner, P. A. M. (2005) Predicting bryophyte diversity in grassland and eucalypt-dominated remnants in subhumid Tasmania. *Journal of Biogeography*, **32**, 2015–2024.

Pirozynski, K. A. & Malloch, D. W. (1975) The origin of land plants: A matter of mycotropism. *Biosystems*, **6**, 153–164.

Pohjamo, M. & Laaka-Lindberg, S. (2003) Reproductive modes in a leafy hepatic *Anastrophyllum hellerianum*. *Perspectives in Plant Ecology, Evolution and Systematics*, **6**, 159–168.

Porley, R. D. & Ellis, R. W. (2002) *Timmia megapolitana* Hedw. (Bryopsida, Timmiales) new to the British Isles. *Journal of Bryology*, **24**, 151–156.

Pressel, S., Ligrone, R., Duckett, J. G. & Davis, E. C. (2008a) A novel ascomycetous endophytic association in the rhizoids of the leafy liverwort family Schistochilaceae (Jungermanniidae, Hepaticopsida). *American Journal of Botany*, **95**, 531–541.

Pressel, S., Ligrone, R. & Duckett, J. G. (2008b) Cellular differentiation in moss protonemata; a morphological and experimental study. *Annals of Botany*, **102**, 227–245.

Price, J. S., Heathwaite, A. L. & Baird, A. J. (2003) Hydrological processes in abandoned and restored peatlands: an overview of management approaches. *Wetlands Ecology and Management*, **11**, 65–83.

Primack, R. B. (1993) *Essentials of Conservation Biology*. Sunderland: Sinauer Associates.

Prins, H. H. T. (1982) Why are mosses eaten in cold environments only. *Oikos*, **38**, 374–380.

Proctor, V. W. (1961) Dispersal of *Riella* spores by waterfowl. *Bryologist*, **64**, 58–61.

Proctor, M. C. F. (2000) The bryophyte paradox: tolerance of desiccation, evasion of drought. *Plant Ecology*, **151**, 41–49.

Proctor, M. C. F. (2001) Patterns of desiccation tolerance and recovery in bryophytes. *Plant Growth Regulation*, **35**, 147–156.

Proctor, M. C. F. (2003) Experiments on the effect of different intensities of desiccation on bryophyte survival, using chlorophyll fluorescence as an index of recovery. *Journal of Bryology*, **25**, 201–210.

Proctor, M. C. F. (2004) Light and desiccation responses of *Weymouthia mollis* and *W. cochlearifolia*, two pendulous rainforest epiphytes from Australia and New Zealand. *Journal of Bryology*, **26**, 167–173.

Proctor, M. C. F. (2005) Why do Polytrichaceae have lamellae? *Journal of Bryology*, **27**, 221–230.

Proctor, M. C. F. (2009) Physiological ecology. In *Bryophyte Biology*, 2nd edn, eds. B. Goffinet & A. J. Shaw. Cambridge: Cambridge University Press, pp. 237–268.

Proctor, M. C. F., Oliver, M. J., Wood, A. J., *et al.* (2007) Desiccation-tolerance in bryophytes: a review. *Bryologist*, **110**, 595–621.

Pypker, T. G., Unsworth, M. H. & Bond, B. J. (2006a) The role of epiphytes in rainfall interception by forests in the Pacific Northwest. I. Laboratory measurements of water storage. *Canadian Journal of Forest Research*, **36**, 809–818.

Pypker, T. G., Unsworth, M. H. & Bond, B. J. (2006b) The role of epiphytes in rainfall interception by forests in the Pacific Northwest. II. Field measurements at the branch and canopy scale. *Canadian Journal of Forest Research*, **36**, 819–832.

Qian, H. (1999) Spatial pattern of vascular plant diversity in North America North of Mexico and its floristic relationships with Eurasia. *Annals of Botany*, **83**, 271–283.

Qiu, Y. L., Li, L., Wang, B., *et al.* (2006) The deepest divergences in land plants inferred from phylogenomic evidence. *Proceedings of the National Academy of Sciences of the USA*, **103**, 15511–15516.

Raghoebarsing, A. A., Smolders, A. J. P., Schmid, M. C., *et al.* (2005) Methanotrophic symbionts provide carbon for photosynthesis in peat bogs. *Nature*, **436**, 1153–1156.

Raven, P. H. & Axelrod, D. L. (1974) Angiosperm biogeography and past continental movements. *Annals of the Missouri Botanical Garden*, **61**, 539–673.

Read, D. J., Duckett, J. G., Francis, R., Ligrone, R. & Russell, A. (2000) Symbiotic fungal associations in 'lower' land plants. *Philosophical Transactions of the Royal Society of London B*, **355**, 815–830.

Redecker, D., Kodner, R. & Graham, L. E. (2000) Glomalean fungi from the Ordovician. *Science*, **289**, 1920–1921.

Reese, W. D. (1981) '*Chlorochytrium*', a green alga endophytic in Musci. *Bryologist*, **84**, 75–78.

Remy, W., Taylor, T. N., Hass, H. & Kerp, H. (1994) Four hundred-million-year-old vesicular arbuscular mycorrhizae. *Proceedings of the National Academy of Sciences of the USA*, **91**, 11841–11843.

Renzaglia, K. S. & McFarland, K. D. (1999) Antheridial plants of *Megaceros aenigmaticus* in the Southern Appalachians: anatomy, ultrastructure and population distribution. *Haussknechtia Beiheft*, **9**, 307–316.

Renzaglia, K. S., Schuette, S., Duff, R. J., *et al.* (2007) Bryophyte phylogeny: advancing the molecular and morphological frontiers. *Bryologist*, **110**, 179–213.

Renzaglia, K. S., Villareal, J. C. & Duff, R. J. (2009) New insights into morphology, anatomy and systematics of hornworts. In *Bryophyte Biology*, 2nd edn, eds. B. Goffinet & A. J. Shaw. Cambridge: Cambridge University Press, pp. 139–171.

Richardson, D. M., Pysek, P., Rejmanek M., *et al.*, (2000) Naturalizations and invasions of alien plants: concepts and definitions. *Diversity and Distributions*, **6**, 93–107.

Rieley, J. O., Richards, P. W. & Bebbington, A. D. L. (1979) The ecological role of bryophytes in a North Wales woodland. *Journal of Ecology*, **67**, 497–527.

Rikkinen, J. & Virtanen, V. (2008) Genetic diversity in cyanobacterial symbionts of thalloids bryophytes. *Journal of Experimental Botany*, **59**, 1013–1021.

Rochefort, L. (2000) *Sphagnum*: a keystone genus in habitat restoration. *Bryologist*, **103**, 503–508.

Rochefort, L. & Lode, E. (2006) Restoration of degraded boreal peatlands. In *Boreal Peatland Ecosystems*, eds. R. K. Wieder & D. H. Vitt. Berlin Heidelberg: Springer Verlag, pp. 381–422.

Rodgers, G. A. & Stewart, D. P. (1977) The cyanophyte–hepatic symbiosis. I. Morphology and physiology. *New Phytologist*, **78**, 441–458.

Rothero, G. P., Duckett, J. G. & Pressel, S. (2006) Active conservation: augmenting the only British population of *Bryum schleicheri* var. *latifolium* via in vitro cultivation. *Field Bryology*, **90**, 12–16.

Roy, S., Pellinen, J., Sen, C. K. & Hänninen, O. (1995) Benzo(a) anthracene and benzo(a) pyrene exposure in the aquatic plant *Fontinalis antipyretica*: uptake, elimination and the response to biotransformation and antioxidant enzymes. *Chemosphere*, **29**, 1301–1311.

Rozzi, R., Armesto, J. J., Goffinet, B., *et al.* (2008) Changing lenses to assess biodiversity: patterns of species richness in sub-Antarctic plants and implications for global conservation. *Frontiers in Ecology and The Environment*, **6**, 131–137.

Russell, J. & Bulman, S. (2005) The liverwort *Marchantia foliacea* forms a specialized symbiosis with AM fungi in the genus *Glomus*. *New Phytologist*, **165**, 567–579.

Rütten, D. & Santarius, K. A. (1992) Relationship between frost tolerance and sugar concentration of various bryophytes in summer and winter. *Oecologia*, **91**, 260–265.

Rütten, D. & Santarius, K. A. (1993) Seasonal variation in frost tolerance and sugar content of two *Plagiomnium* species. *Bryologist*, **96**, 564–568.

Rydgren, K. & Økland, R. (2002) Sex distribution and sporophyte frequency in a population of the clonal moss *Hylocomium splendens*. *Journal of Bryology*, **24**, 207–214.

Rydgren, K. & Økland, R. (2003) Short-term costs of sexual reproduction in the clonal moss *Hylocomium splendens*. *Bryologist*, **106**, 212–220.

Rydgren, K., Cronberg, N. & Økland, R. H. (2006) Factors influencing reproductive success in the clonal moss, *Hylocomium splendens*. *Oecologia*, **147**, 445–454.

Rydin, H. (2009) Population and community ecology of bryophytes. In *Bryophyte Biology*, 2nd edn, eds. A. J. Shaw & B. Goffinet. Cambridge: Cambridge University Press, pp. 393–444.

Rydin, H. & Jeglum, J. K. (2006) *The Biology of Peatlands*. Oxford: Oxford University Press.

Sagot, C. & Rochefort, L. (1996) Tolérance des sphaignes à la dessication. *Cryptogamie, Bryologie Lichénologie*, **17**, 171–183.

Salonen, V. & Setälä, H. (1992) Plant colonization of bare peat surface: relative importance of seed availability and soil. *Ecography*, **15**, 199–204.

Salonen, V., Penttinen, A. & Särkkä, A. (1992) Plant colonization of a bare peat surface: population changes and spatial patterns. *Journal of Vegetation Science*, **3**, 113–118.

Sanderson, M. J. (2003) Molecular data from 27 proteins do not support a Precambrian origin of land plants. *American Journal of Botany*, **90**, 954–956.

Sanmartin, I., Wanntorp, L. & Winkworth, R. C. (2007) West Wind Drift revisited: testing for directional dispersal in the Southern Hemisphere using event-based tree fitting. *Journal of Biogeography*, **34**, 398–416.

Sardans, J. & Peñuelas, J. (2008) Drought changes nutrient sources, content and stoichiometry in the bryophyte *Hypnum cupressiforme* Hedw. growing in a Mediterranean forest. *Journal of Bryology*, **30**, 59–65.

Såstad, S. M., Pedersen, B. & Digre, K. (1999) Habitat-specific genetic effects on growth rate and morphology across pH and water level gradients within a population of the moss *Sphagnum angustifolium* (Sphagnaceae). *American Journal of Botany*, **86**, 1687–1698.

Satake, K., Nishikawa, M. & Shibata, K. (1989) Distribution of aquatic bryophytes in relation to water chemistry of the acid river Akagawa, Japan. *Archiv für Hydrobiologie*, **116**, 299–311.

Schnittler, M. (2001) Foliicolous liverworts as a microhabitat for Neotropical Myxomycetes. *Nova Hedwigia*, **72**, 259–270.

Schofield, W. B. (1984) Bryogeography of the Pacific coast of North America. *Journal of the Hattori Botanical Laboratory*, **55**, 35–43.

Schofield, W. B. (1988) Bryophyte disjunctions in the northern hemisphere: Europe and North America. *Botanical Journal of the Linnean Society*, **98**, 211–224.

Schofield, W. B. (1992) Bryophyte distribution patterns. In *Bryophytes and Lichens in a Changing World*, eds. J. W. Bates & A. M. Farmer. Oxford: Clarendon Press, pp. 103–130.

Schrenk, C., Pflugmacher, S., Brüggemann, R., *et al.* (1998) Glutathione S-transferase activity in aquatic macrophytes with emphasis on habitat dependence. *Ecotoxicology and Environmental Safety*, **40**, 226–233.

Schulze, C. H., Waltert, M., Kessler, P. J. A., *et al.* (2004) Biodiversity indicator groups of tropical land-use systems: comparing plants, birds, and insects. *Ecological Applications*, **14**, 1321–1333.

Schüßler, A. (2000) *Glomus claroideum* forms an arbuscular-mycorrhiza-like symbiosis with the hornwort *Anthoceros punctatus*. *Mycorrhiza*, **10**, 15–21.

Schuster, R. M. (1966–1992) *The Hepaticae and Anthocerotae of North America*. Chicago: Field Museum of Natural History.

Schuster, R. M. (1983) Phytogeography of the Bryophyta. In *New Manual of Bryology*, Vol. 1, ed. R. M. Schuster, Nichinan: Hattori Botanical Laboratory, pp. 463–626.

Schuster, R. M. (1984a) Comparative morphology and anatomy of the Hepaticae. In *New Manual of Bryology*, Vol. 2, ed. R. M. Schuster. Nichinan: Hattori Botanical Laboratory, pp. 760–891.

Schuster, R. M. (1984b) Evolution, phylogeny and classification of the Hepaticae. In *New Manual of Bryology*, Vol. 2, ed. R. M. Schuster. Nichinan: Hattori Botanical Laboratory, pp. 892–1069.

Schuster, R. M. (1984c) Morphology, phylogeny and classification of the Anthocerotae. In *New Manual of Bryology*, Vol. 2, ed. R. M. Schuster. Nichinan: Hattori Botanical Laboratory, pp. 1070–1092.

Selosse, M. A. (2005) Are liverworts imitating mycorrhizas? *New Phytologist*, **165**, 345–349.

Selosse, M. A., Baudoin, E. & Vandenkoornhuyse, P. (2004) Symbiotic microorganisms, a key for ecological success and protection of plants. *Comptes Rendus Biologie*, **327**, 639–648.

Sérgio, C., Araujo, M. & Draper, D. (2000) Portuguese bryophyte diversity and priority areas for conservation. *Lindbergia*, **25**, 116–123.

Shaw, A. J. (1987) Evolution of heavy metal tolerance in bryophytes II. An ecological and experimental investigation of the 'copper moss', *Scopelophila cataractae* (Pottiaceae). *American Journal of Botany*, **74**, 813–821.

Shaw, A. J. (1988) Genetic variation for tolerance to copper and zinc within and among populations of the moss, *Funaria hygrometrica* Hedw. *New Phytologist*, **109**, 211–222.

Shaw, A. J. (1992) The evolutionary capacity of bryophytes and lichens. In *Bryophytes and Lichens in a Changing Environment*, eds. J. W. Bates & A. M. Farmer. Oxford: Clarendon Press, pp. 362–380.

Shaw, A. J. (1994) Adaptation to metals in widespread and endemic plants. *Environmental Health Perspectives*, **102**, Suppl. 12, 105–108.

Shaw, A. J. (2000) Molecular phylogeography and cryptic speciation in the mosses, *Mielichhoferia elongata* and *M. mielichhoferiana* (Bryaceae). *Molecular Ecology*, **9**, 595–608.

Shaw, A. J. (2001) Biogeographic patterns and cryptic speciation in bryophytes. *Journal of Biogeography*, **28**, 253–261.

Shaw, A. J. (2009) Bryophyte species and speciation. In *Bryophyte Biology*, 2nd edn, eds. A. J. Shaw & B. Goffinet. Cambridge: Cambridge University Press, pp. 445–485.

Shaw, A. J. & Albright, D. (1990) Potential for the evolution of heavy metal tolerance in *Bryum argenteum*, a moss. II. Generalized tolerances among diverse populations. *Bryologist*, **93**, 187–192.

Shaw, A. J. & Owens, H. (1995) Ecological and experimental studies on the 'copper mosses': *Mielichhoferia elongata* (Bryaceae) and *Scopelophila cataractae* (Pottiaceae). *Fragmenta Floristica et Geobotanica*, **40**, 519–531.

Shaw, A. J., Meagher, T. R. & Harley, P. (1987) Electrophoretic evidence of reproductive isolation between two varieties of the moss, *Climacium americanum*. *Heredity*, **59**, 337–343.

Shaw, A. J., Beer, S. C. & Lutz, J. (1989) Potential for the evolution of heavy metal tolerance in *Bryum argenteum*, a moss. I. Variation within and among populations. *Bryologist*, **92**, 72–80.

Shaw, A. J., Werner, O. & Ros, R. M. (2003) Intercontinental Mediterranean disjunct mosses: morphological and molecular patterns. *American Journal of Botany*, **90**, 540–550.

Shaw, A. J., Cox, C. J. & Goffinet, B. (2005) Global patterns of moss diversity: taxonomic and molecular inferences. *Taxon*, **54**, 337–352.

Shimamura, M., Yamaguchi, T. & Deguchi, H. (2008) Airborne sperm of *Conocephalum conicum* (Conocephalaceae). *Journal of Plant Research*, **121**, 69–71.

Sillett, S. C., Gradstein, S. R. & Griffin, D. (1995) Bryophyte diversity of *Ficus* tree crowns from cloud forest and pasture in Costa Rica. *Bryologist*, **98**, 251–260.

Simard, M., Lecomte, N., Bergeron, Y., Bernier, P. & Paré, D. (2007) Forest productivity decline by successional paludification of boreal soils. *Ecological Applications*, **17**, 1619–1637.

Slack, N. G. (1990) Bryophytes and ecological niche theory. *Botanical Journal of the Linnean Society*, **104**, 187–213.

Smith, R. M., Young, M. R. & Marquiss, M. (2001) Bryophyte use by an insect herbivore: does the cranefly *Tipula montana* select food to maximize growth? *Ecological Entomology*, **26**, 83–90.

Snäll, T., Fogelqvist, J., Bibeiro, P. J. & Lascoux, L. (2004a) Spatial genetic structure in two congeneric epiphytes with different dispersal strategies analysed by three different methods. *Molecular Ecology*, **13**, 2109–2119.

Snäll, T., Hagstrom, A., Rudolphi, J. & Rydin, H. (2004b) Distribution pattern of the epiphyte *Neckera pennata* on three spatial scales: importance of past landscape structure, connectivity and local conditions. *Ecography*, **27**, 757–766.

Söderström, L. (1992) Invasions and range expansions and contractions of bryophytes. In *Bryophytes and Lichens in a Changing Environment*, eds. J. W. Bates & A. M. Farmer. Oxford: Clarendon Press, pp. 131–158.

Söderström, L. & Herben, T. (1997) Dynamics of bryophyte metapopulations. *Advances in Bryology*, **6**, 205–240.

Söderström, L., Séneca, A. & Santos, M. (2007) Rarity patterns in the northern hemisphere members of the Lophoziaceae/Scapaniaceae complex (Hepaticae, Bryophyta). *Biological Conservation*, **135**, 352–359.

Solheim, B. & Zielke, M. (2002) Associations between cyanobacteria and mosses. In *Cyanobacteria in Symbiosis*, eds. A. N. Rai, B. Bergman & U. Rasmussen. Dordrecht: Kluwer, pp. 137–152.

Sotiaux, A., Pioli, A., Royaud, A., Schumacker, R. & Vanderpoorten, A. (2007) A checklist of the bryophytes of Corsica (France): new records and a review of the literature. *Journal of Bryology*, **29**, 41–53.

Stark, L. R. (2005) Phenology of patch hydration, patch temperature and sexual reproductive output over a four-year period in the desert moss *Crossidium crassinerve*. *Journal of Bryology*, **27**, 231–240.

Stark, L. R. & McLetchie, D. N. (2006) Gender-specific heat-shock tolerance of hydrated leaves in the desert moss *Syntrichia caninervis*. *Physiologia Plantarum*, **126**, 187–195.

Stark, L. R., Mishler, B. D. & McLetchie, D. N. (2000) The cost of realized sexual reproduction: assessing patterns of reproductive allocation and sporophyte abortion in a desert moss. *American Journal of Botany*, **87**, 1599–1608.

Steinman, A. D. (1994) The influence of phosphorus enrichment on lotic bryophytes. *Freshwater Biology*, **31**, 53–63.

Stenoien, H. K. (2008) Slow molecular evolution in 18S rDNA, *rbc*L and *nad*5 genes of mosses compared with higher plants. *Journal of Evolutionary Biology*, **21**, 566–571.

Stephenson, S. L., Studlar, S. M., McQuattie, C. J. & Edwards, P. J. (1995) Effects of acidification on bryophyte communities in West Virginia mountain streams. *Journal of Environment Quality*, **4**, 116–124.

Stieperaere, H. (1994) *Lophocolea semiteres* (Lehm.) Mitt. In Belgium and The Netherlands, another antipodal bryophyte spreading on the European continent. *Lindbergia*, **19**, 29–36.

Sundberg, S. & Rydin, H. (2002) Habitat requirements for establishment of *Sphagnum* by spores. *Journal of Ecology*, **90**, 268–278.

Suren, A. M. (1991) Bryophytes as invertebrate habitat in two New Zealand alpine streams. *Freshwater Biology*, **26**, 399–418.

Szövényi, P., Hock, Z. & Tóth, Z. (2004) Phorophyte preference of epiphytic bryophytes in a stream valley in the Carpathian Basin. *Journal of Bryology*, **26**, 137–146.

Szövényi, P., Hock, Z., Urmi, E. & Schneller, J. J. (2006) Contrasting phylogeographic patterns in *Sphagnum fimbriatum* and *Sphagnum squarrosum* (Bryophyta, Sphagnopsida) in Europe. *New Phytologist*, **172**, 784–794.

Szövényi, P., Hock, Z., Schneller, J. J. & Tóth, Z. (2007) Multilocus dataset reveals demographic histories of two peat mosses in Europe. *BMC Evolutionary Biology*, **7**, art. 144.

Takhtajan, A. L. (1986) *The Floristic Regions of the World*. Berkeley: UC Press.

Tallis, J. H. (1964) Studies on southern Pennines. III. The behaviour of *Sphagnum*. *Journal of Ecology*, **52**, 345–353.

Tallis, J. H. (1995) Climate and erosion signals in British blanket peats: the significance of *Racomitrium lanuginosum* remains. *Journal of Ecology*, **83**, 1021–1030.

Tan, B. C. & Pocs, T. (2000) Bryogeography and conservation of bryophytes. In *Bryophyte Biology*, 1st edn, eds. A. J. Shaw & B. Goffinet. Cambridge: Cambridge University Press, pp. 403–448.

Tanahashi, T., Sumikawa, N., Kato, M. & Hasebe, M. (2005) Diversification of gene function: homologs of the floral regulator FLO/LFY control the first zygotic cell division in the moss *Physcomitrella patens. Development*, **132**, 1727–1736.

Taylor, T. N., Klavins, S. D., Krings, M., *et al.* (2004) Fungi from the Rhynie chert: a view from the dark side. *Transactions of the Royal Society of Edinburgh: Earth Sciences*, **94**, 457–473.

Thiébaut, G., Vanderpoorten, A., Guérold, F., Boudot, J.-P. & Muller, S. (1998) Bryological patterns and streamwater acidification in the Vosges Mountains (N.E. France): an analysis tool for the survey of acidification processes. *Chemosphere*, **36**, 1275–1289.

Thingsgaard, K. (2001) Population structure and genetic diversity of the amphiatlantic haploid peatmoss *Sphagnum affine* (Sphagnopsida). *Heredity*, **87**, 485–496.

Tivy, J. (1993) *Biogeography. A Study of Plants in the Ecosphere*, 3rd edn. Harlow: Longman group.

Tremp, H. & Kohler, A. (1993) *Wassermoose als Versauerungsindikatoren. Praxiesorientierte Bioindikationsvervahren mit Wassermoosen zur Überwachung des Zauerezustandes von pufferschwachen Fliessgewässern*. Karlsruhe: Landesanstalt für Umweltschutz Baden-Württemberg.

Tuba, Z., Csintalan, Z. & Proctor, M. C. F. (1996) Photosynthetic responses of a moss, *Tortula ruralis* ssp. *ruralis*, and the lichens *Cladonia convoluta* and *C. furcata* to water deficit and short periods of dessication, and their ecophysiological significance: a baseline study at present-day CO_2 concentrations. *New Phytologist*, **133**, 353–361.

Turetsky, M. R. & St Louis, V. L. (2006) Disturbance in boreal peatlands. In *Boreal Peatland Ecosystems*, eds. R. K. Wieder & D. H. Vitt. Berlin-Heidelberg: Springer, pp. 359–379.

Ueno, T., Bekku, Y., Uchida, M. & Kanda, H. (2006) Photosynthetic light responses of a widespread moss, *Sanionia uncinata*, from contrasting water regimes in the high Arctic tundra, Svalbard, Norway. *Journal of Bryology*, **28**, 345–349.

Upchurch, P. (2008) Gondwanan break-up: legacies of a lost world? *Trends in Ecology and Evolution*, **23**, 229–236.

VanAller Hernick, L., Landing, E. & Bartowski, K. E. (2008) Earth's oldest liverworts: *Metzgeriothallus sharonae* sp. nov. from the Middle Devonian (Givetian) of eastern New York, USA. *Review of Paleobotany and Palynology*, **148**, 154–162.

van Breemen, N. (1995) How *Sphagnum* bogs down other plants. *Trends in Ecology and Evolution*, **10**, 270–275.

Vanderpoorten, A. (1999) Aquatic bryophytes for a spatio-temporal monitoring of the water pollution of the rivers Meuse and Sambre (Belgium). *Environmental Pollution*, **104**, 401–410.

Vanderpoorten, A. (2003) Hydrochemical determinism, ecological polymorphism and indicator values of aquatic bryophytes for water quality. In *Modern Trends in Applied Aquatic Ecology*, eds. R. S. Ambasht & N. K. Ambasht. New York: Kluwer Academic Publishers, pp. 65–96.

Vanderpoorten, A. (2004) A simple treatment for a complicated evolutionary history: the genus *Hygroamblystegium* (Hypnales, Amblystegiaceae). *Monographs in Systematic Botany from the Missouri Botanical Garden*, **98**, 320–327.

Vanderpoorten, A. & Engels, P. (2002) The effects of environmental variation on bryophytes at a regional scale. *Ecography*, **25**, 513–522.

Vanderpoorten, A. & Engels, P. (2003) Patterns of bryophyte diversity and rarity at a regional scale. *Biodiversity and Conservation*, **12**, 545–553.

Vanderpoorten, A. & Hallingbäck, T. (2009) Conservation biology. In *Bryophyte Biology*, 2nd edn, eds. B. Goffinet & A. J. Shaw. Cambridge: Cambridge University Press, pp. 487–533.

Vanderpoorten, A., Sotiaux, A. & Engels, P. (2004a) Trends in diversity and abundance of obligate epiphytic bryophytes in a highly managed landscape. *Ecography*, **27**, 567–576.

Vanderpoorten, A., Delescaille, L. & Jacquemart, A.-L. (2004b) The bryophyte layer in a calcareous grassland after a decade of contrasting mowing regimes. *Biological Conservation*, **117**, 11–18.

Vanderpoorten, A., Carine, M. A. & Rumsey, F. (2007) Does Macaronesia exist? Conflicting signals in the bryophyte and pteridophyte floras. *American Journal of Botany*, **94**, 625–639.

van der Velde, M. & Bijlsma, R. (2000) Amount and structure of intra- and interspecific genetic variation in the moss genus *Polytrichum*. *Heredity*, **85**, 328–337.

van der Wal, R. & Brooker, R. W. (2004) Mosses mediate grazer impacts on grass abundance in arctic ecosystems. *Functional Ecology*, **18**, 77–86.

van der Wijk, R., Margadant, W. D. & Florschütz, P. A. (1959) Index Muscorum. *Regnum Vegetabile* **17**.

van Tooren, B. F., Odé, B., During, H. J. & Bobbink, R. (1990) Regeneration of species richness in the bryophyte layer of Dutch chalk grasslands. *Lindbergia*, **16**, 153–160.

van Tooren, B. F., Odé, B. & Bobbink, R. (1991) Management of Dutch chalk grasslands and the species richness of the cryptogam layer. *Acta Botanica Neerlandica*, **40**, 379–380.

van Zanten, B. O. (1978) Experimental studies on transoceanic long-range dispersal of moss spores in the southern hemiphere. *Journal of the Hattori Botanical Laboratory*, **44**, 455–482.

Villarreal, J. C. & Renzaglia, K. S. (2006) Structure and development of *Nostoc* strands in *Leiosporoceros dussii* (Anthocerotophyta): a novel symbiosis in land plants. *American Journal of Botany*, **93**, 693–705.

Virtanen, V., Korpelainen, H. & Kostamo, K. (2007) Forensic botany: usability of bryophyte material in forensic studies. *Forensic Science International*, **172**, 161–163.

Vitt, D. H. (1981) Adaptive modes of the moss sporophyte. *Bryologist*, **84**, 166–186.

Vitt, D. H. (1984) Classification of the Bryopsida. In *New Manual of Bryology*, Vol. 2, ed. R. M. Schuster, Nichinan: Hattori Botanical Laboratory, pp. 696–759.

Vitt, D. H. & Wieder, R. K. (2009) The structure and function of bryophyte dominated peatlands. In *Bryophyte Biology*. 2nd edn, eds. B. Goffinet & A. J. Shaw. Cambridge: Cambridge University Press, pp. 357–391.

Voth, P. D. (1943) Effects of nutrient solution concentration on the growth of *Marchantia polymorpha*. *Botanical Gazette*, **104**, 591–601.

Wall, D. P. (2005) Origin and rapid diversification of a tropical moss. *Evolution*, **59**, 1413–1424.

Wang, B. & Qiu, Y.-L. (2006) Phylogenetic distribution and evolution of mycorrhizas in land plants. *Mycorrhiza*, **16**, 299–363.

Wasley, J., Robinson, S. A., Lovelock, C. E. & Popp, M. (2006) Climate change manipulations show Antarctic flora is more strongly affected by elevated nutrients than water. *Global Change Biology*, **12**, 1800–1812.

Welch, J. J. & Bromham, L. (2005) Molecular dating when rates vary. *Trends in Ecology and Evolution*, **20**, 320–327.

Wellman, C. H. & Gray, J. (2000) The microfossil record of early land plants. *Philosophical Transactions of the Royal Society of London B*, **355**, 717–732.

Wells, J. M. & Brown, D. H. (1996) Mineral nutrient recycling within shoots of the moss *Rhytidiadelphus squarrosus* in relation to growth. *Journal of Bryology*, **19**, 1–17.

Whinam, J. & Buxton, R. (1997) *Sphagnum* peatlands of Australia: an assessment of harvesting sustainability. *Biological Conservation*, **82**, 21–29.

Whinam, J., Barmuta, L. A. & Chilcott, N. (2001) Floristic description and environmental relationships of Tasmanian *Sphagnum* communities and their conservation management. *Australian Journal of Botany*, **49**, 673–685.

Whittaker, R. J. (1998) *Island Biogeography: Ecology, Evolution and Conservation.* Oxford: Oxford University Press.

Wigginton, M. J. (2004) Checklist and distribution of the liverworts and hornworts of sub-Saharan Africa, including the East African Islands. *Tropical Bryology Research Report*, **5**, 1–102.

Wiklund, K. & Rydin, H. (2004a) Ecophysiological constraints on spore establishment in bryophytes. *Functional Ecology*, **18**, 907–913.

Wiklund, K. & Rydin, H. (2004b) Colony expansion of *Neckera pennata*: Modelled growth rate and effect of microhabitat, competition, and precipitation. *Bryologist*, **107**, 293–301.

Williams, C. B. & Sillett, S. C. (2007) Epiphyte communities on redwood (*Sequoia sempervirens*) in northwestern California. *Bryologist*, **110**, 420–452.

Wilson, P. J. & Provan, J. (2003) Effect of habitat fragmentation on levels and patterns of genetic diversity in natural populations of the peat moss *Polytrichum commune*. *Proceedings of the Royal Society of London B*, **270**, 881–886.

Wolf, J. H. D. (1993) Diversity patterns and biomass of epiphytic bryophytes and lichens along an altitudinal gradient in the northern Andes. *Annals of the Missouri Botanical Garden*, **80**, 928–960.

Woolgrove, C. E. & Woodin, S. J. (1996) Current and historical relationships betweeen the tissue nitrogen content of a snowbed bryophyte and nitrogenous air pollution. *Environmental Pollution*, **91**, 283–288.

Yang, R.-D., Mao, J.-R., Zhang, W.-H., Jiang, L.-J. & Gao, H. (2004) Bryophyte-like fossil (*Parafunaria sinensis*) from Early-Middle Cambrian Kaili formation in Guizhou Province, China. *Acta Botanica Sinica*, **46**, 180–185.

Yates, C. J., Norton, D. A. & Hobbs, R. J. (2000) Grazing effects on plant cover, soil and microclimate in fragmented woodlands in south-western Australia: implications for restoration. *Austral Ecology*, **25**, 36–47.

Zamfir, M. (2000) Effects of bryophytes and lichens on seedling emergence of alvar plants: evidence from greenhouse experiments. *Oikos*, **88**, 603–611.

Zartman, C. E. & Nascimento, H. E. M. (2006) Are habitat-tracking metacommunities dispersal-limited? Inferences from abundance–occupancy patterns of epiphylls in Amazonian forest fragments. *Biological Conservation*, **127**, 46–54.

Zartman, C. E. & Shaw, A. J. (2006) Metapopulation extinction thresholds in rain forest remnants. *American Naturalist*, **167**, 177–189.

Zechmeister, H. (2005) Bryophytes of continental salt meadows in Austria. *Journal of Bryology*, **27**, 297–302.

Zechmeister, H. G. & Hohenwallner, D. (2006) A comparison of biomonitoring methods for the estimation of atmospheric pollutants in an industrial town in Austria. *Environmental Monitoring and Assessment*, **117**, 245–259.

Zechmeister, H. G., Schmitzberger, I., Steurer, B., Peterseil, J. & Wrbka, T. (2003a) The influence of land-use practices and economics on plant species richness in meadows. *Biological Conservation*, **114**, 165–177.

Zechmeister, H. G., Grodzinska, K. & Szarek-Lukaszewska, G. (2003b) Bryophytes. In *Bioindicators and Biomonitors. Principles, Concepts and Applications*, eds. B. A. Markert, A. M. Breure & H. G. Zechmeister. Oxford: Elsevier, pp. 329–375.

Zechmeister, H., Dullinger, S., Hohenwallner, D., *et al.* (2006) Pilot study on road traffic emissions (PAHs, heavy metals) measured by using mosses in a tunnel experiment in Vienna, Austria. *Environmental Science Pollution Research*, **13**, 398–405.

Zechmeister, H., Moser, D. & Milasowszky, N. (2007a) Spatial distribution patterns of *Rhynchostegium megapolitanum* at the landscape scale: an expanding species? *Applied Vegetation Science*, **10**, 111–120.

Zechmeister, H., Dirnböck, T., Hülber, K. & Mirtl, M. (2007b) Assessing airborne pollution effects on bryophytes: lessons learned through long-term integrated monitoring in Austria. *Environmental Pollution*, **147**, 696–705.

Zhang, Y. & Guo, L.-D. (2007) Arbuscular mycorrhizal structure and fungi associated with mosses. *Mycorrhiza*, **17**, 319–325.

Index

Entries in bold refer to figures